中国植物病理学会第十三届青年学术研讨会论文选编

植物病理学研究进展

吴学宏　主编

中国农业大学出版社
·北京·

内容简介

本书是中国植物病理学会第十三届青年学术研讨会论文选编，共收录了96篇研究论文、简报和摘要。涉及植物病原菌物、病毒、原核生物、线虫及其所致病害，包括病原鉴定和检测、病原一寄主植物互作、病害流行及预测、抗病性及抗病育种、生物防治、种子病理与杀菌剂、病害综合防治、分子生物学及其应用、信息技术及其应用等方面，基本上反映了近两年来我国植物病理学青年科技工作者在植物病理学各分支学科基础理论、应用基础和植物病害防治实践等方面所取得的研究进展。

本书可供植物病理学、农学、生物学等相关专业教师、本科生和研究生、科研人员、技术推广人员以及有关管理人员参考。

图书在版编目(CIP)数据

植物病理学研究进展 / 吴学宏主编．—北京：中国农业大学出版社，2017.11
ISBN 978-7-5655-1932-1

Ⅰ.①植…　Ⅱ.①吴…　Ⅲ.①植物病理学-文集　Ⅳ.①S432.1-53

中国版本图书馆 CIP 数据核字(2017)第 260441 号

书　名　植物病理学研究进展
作　者　吴学宏　主编

策划编辑　赵　中　　**责任编辑**　田树君
封面设计　郑　川
出版发行　中国农业大学出版社
社　址　北京市海淀区圆明园西路 2 号　　**邮政编码**　100193
电　话　发行部 010-62818525，8625　　读者服务部 010-62732336
编辑部 010-62732617，2618　　出　版　部 010-62733440
网　址　http://www.caupress.cn　　**E-mail** cbsszs @ cau.edu.cn
经　销　新华书店
印　刷　涿州市星河印刷有限公司
版　次　2017 年 11 月第 1 版　2017 年 11 月第 1 次印刷
规　格　787×1 092　16 开本　9.5 印张　240 千字
定　价　30.00 元

图书如有质量问题本社发行部负责调换

中国植物病理学会第十三届青年学术研讨会
组织委员会

顾　　问：彭友良

主　　任：吴学宏

副主任：王海光　燕继晔　杨　俊　左示敏

委　　员：（按姓氏笔画为序）

王　勇　王海光　王晓杰　左示敏　刘大伟　刘文德
刘永锋　吴学宏　李　韬　李　燕　杨　俊　汪　涛
陈夕军　陈旭君　陈孝仁　周　涛　欧阳寿强
曹志艳　梁春浩　谢甲涛　窦道龙　燕继晔　魏太云

前　言

中国植物病理学会第十三届青年学术研讨会论文选编《植物病理学研究进展》共收录了96篇研究论文、简报和摘要。内容涉及植物病原菌物、病毒、原核生物、线虫等及其所致病害，包括病原鉴定和检测、病原—寄主植物互作、病害流行及预测、抗病性及抗病育种、生物防治、种子病理与杀菌剂、病害综合防治、分子生物学及其应用、信息技术及其应用等方面，基本上反映了近两年来我国植物病理学青年工作者在植物病理学各分支学科基础理论、应用基础和植物病害防治实践等方面所取得的研究进展。

本次大会由中国植物病理学会青年委员会主办，由扬州大学、江苏省植物病理学会、江苏省农业科学院植物保护研究所、扬州大学教育部植物功能基因组学重点实验室、江苏省粮食作物现代产业技术协同创新中心（培育）、扬州大学教育部农业与农产品安全国际合作联合实验室共同承办，以“青年植病研究与协同创新”为会议主题。大会的召开和论文集的出版得到了中国科学技术协会和中国植物病理学会的资助，挂靠单位中国农业大学给予了关心和指导，我国植物病理学界多位专家、教授给予了关心、指导和帮助，大会承办单位和编委会的同志们付出了辛苦的劳动，中国农业大学出版社给予了大力支持和帮助。在此，我们表示衷心的感谢！

由于大会征集论文时间紧，同时，本着尊重作者意愿和文责自负的原则，编辑过程中，对论文内容一般未作改动，仅在编辑体例上进行了处理。因此，错误和不足之处在所难免，敬希论文作者和读者批评指正！

编　者

2017 年 10 月

目　录

In vitro replication system mediated by NIb of *Wheat yellow mosaic virus*

Guowei Geng, Chengming Yu, Xiangdong Li, Xuefeng Yuan*

(*Department of Plant Pathology, College of Plant Protection, Shandong Agricultural University, Shandong Province Key Laboratory of Agricultural Microbiology, Tai'an 271018, China*)

Abstract: This study was focused on the construction of in vitro replication system mediated by NIb of *Wheat yellow mosaic virus* (WYMV). NIb coding sequences of WYMV were amplified and inserted into the basic vector pMAL-C2X to construct the prokaryotic expression plasmid pMAL-WY-NIb. NIb fused with MBP-tag were expressed with the molecular weight of 100 kD under the induction of 0.4 mM IPTG. Based on the temperature courses assay, it is suggested that 20℃ is optimal and the soluble ratio of MBP-NIb is about 30%, which is enough for affinity chromatography against MBP-tag. Through comparing with other proteins of WYMV and another viral RdRp, the purified MBP-NIb can specifically recognize the 3′ terminal region of WYMV RNA1 or RNA2 and perform *in vitro* replication. This *in vitro* replication system will facilitate the study of mechanism on WYMV replication.

Keywords: *Wheat yellow mosaic virus*; *In vitro* replication; NIb; RNA-dependent RNA polymerase

1 Introduction

In vivo replication of RNA viruses is accomplished by multi-subunit replicase complexes, which consist of viral and host proteins (Buck 1996; Lai 1998; Nagy 2015). The core catalytic subunit of the replicase complexes is a RNA-dependent RNA polymerase (RdRp) encoded by RNA viruses, which specifically interacts with the 3′-end of the viral template facilitating the replication of complementary strand (te Velthuis 2014; Hodge et al., 2016). For identifying the detailed mechanism how RdRp perform the replication of RNA viruses, *in vitro* replication systems specific to given RNA viruses have been set up using purified RdRp from viral-infected cells or heterologous expression systems (Hong et al., 1996; Zhong et al., 2000; Panavas et al., 2002; Rajendran et al., 2002).

Wheat yellow mosaic virus (WYMV) is the main pathogen of soil-borne wheat mosaic

Corresponding author: Xuefeng Yuan; E-mail: snowpeak77@163.com.

disease, which frequently occurred in Japan, China, South Korea of East Asia and seriously threaten the growth and development of wheat (Sawada, 1927; Inouye, 1969; Tao et al., 1980; Wang et al., 1980; Yu et al., 1986; Kuhne 2009; Zhang et al., 2011; Liu et al., 2013). WYMV is naturally transmitted by the *Polymyxa graminis* of the genus *Myxomycetes* producing dormant spore piles, which have strong resistance and can survive 20-25 years in the soil (Abe and Tamada, 1986; Chen 2005). WYMV genome is consist of two linear single-stranded plus-sense RNAs. WYMV RNAs have covalently bound VPg at the 5′end and poly (A) tail at the 3′ end (Han et al., 2000; Ohki et al., 2014). RNA 1 in the length of 7.6 kb and RNA 2 in the length of 3.6 kb respectively encode 270 kDa and 100 kDa polyproteins, which produce functional proteins through the cleavage of proteinase (Li et al., 1997). NIb protein produced from RNA1 has the characteristic of RNA-dependent RNA polymerase (RdRp) and may be involved in the pathogenicity of WYMV (te Velthuis, 2014). In addition, VPg, P3 and 14K may be also associated with *in vivo* accumulation of WYMV based on the data of other similar viruses (Klein et al., 1994; Schaad et al., 1997; Li et al., 1997). In order to identify whether NIb of WYMV is specifically and principally responsible for the replication of WYMV genome and how perform the replication, an *in vitro* replication system need to be built up.

In this study, NIb fused with MBP-tag was expressed from prokaryotic system and purified through affinity chromatography against MBP-tag. Through comparing with other proteins of WYMV and another viral RdRp, the purified MBP-NIb specifically recognized the 3′ region of RNA1 and RNA2 of WYMV and catalyzed *in vitro* synthesis of the complementary strand.

2 Materials and methods

2.1 Construction of prokaryotic expression vectors containing WYMV NIb, VPg, 14K and P3

The primers WYM-R1-4921-BamHI-F and WYM-R1-6504-SalI-R were designed based on WYMV-HC isolate RNA1 (GenBank. AF067124) to amplify the NIb coding sequences of WYMV HC isolate.

The 25 L PCR reaction systems consist of 2.5 μL 10× PCR buffer (Mg^{2+} plus), 1 μL dNTP (each 2.5 mmol/L), 1 μL WYM-R1-4921-BamHI-F and 1 μL WYM-R1-6504-SalI-R, 1 μL pUWR1-1(Dong, 2004), 1 μL Taq DNA polymerase (5 U/μL). Then ddH_2O was added to make a final reaction volume of 25 μL. The PCR reaction process was pre-denaturation at 94℃ for 5 min, followed by 30 cycles reaction including denaturation at 94℃ for 40 s, annealing at 51℃ for 40 s, and extension at 72℃ for 2 min, and a final extension at 72℃ for 10 min.

The PCR products were recovered by DNA recovery kit, digested with BamH I and Sal I, followed by ligation with the same enzymes digested pMAL-C2X (NEB) with T4 DNA ligase (Takara) at 16℃ for at least 8 h. The ligation products were transformed into

E. coli DH5α. The positive clones of pMAL-WY-NIb were identified by colony PCR, restriction enzyme digestion and final DNA sequencing.

Prokaryotic expression vectors containing WYMV VPg, 14K and P3 were constructed as pMAL-WY-NIb.

2.2 Prokaryotic expression of WYMV NIb, VPg, 14K and P3

The prokaryotic expression plasmids containing WYMV NIb, VPg, 14K and P3 were respectively transformed into competent cells of *E. coli* Rossetta and plated with LB plates containing 100 mg/mL ampicillin. The Rossetta strain containing the prokaryotic expression plasmids were identified by colony PCR and then incubated overnight with 200 rpm shaking at 37℃. The overnight culture was added into 30 mL of fresh LB broth at a ratio of 1∶100 and incubated at 37℃ until OD600 to 0.4～0.6. The cultures were placed on ice for 10 min and then distributed to finger tubes at 3 mL/ tube. IPTG was added to make a final concentration of 0 mmol/L, 0.2 mmol/L, 0.4 mmol/L, 0.6 mmol/L, 0.8 mmol/L and 1.0 mmol/L, respectively. All tube were then incubated at 37℃ with 200 rpm for 4 h. 1 mL culture from different tubes were collected at 13 000 r/min for 30 s. 100 μL ddH_2O was added to suspend the supernatant and then an equal volume of 2×SDS protein loading buffer was added. Protein samples were analyzed using SDS-PAGE to detect the expression pattern under the induction of different concentration of IPTG.

2.3 Solubility analysis of prokaryotic expression proteins at different temperature

According to the steps of 2.2, 10 mL culture was incubated at different temperature with the induction of optimal IPTG. Then 1mL cultures were concentrated and resuspended using 1.0 mL of protein buffer (20 mmol/L Tris-HCl (pH 7.4), 0.2 mol/L NaCl, 1 mmol/L EDTA). The ice-cold protein samples were then underwent ultrasonic process at the following condition: 400 W, 100 times with working 4s and interval 4 s. The ultrasonic treated samples were separated into supernatant and sediment by centrifugation with 13 000 r/min for 15 min at 4℃. 100 μL of supernatant was mixed with 100 μL of 2×SDS protein loading buffer. Sediment was suspended using 200 μL of 1×SDS protein loading buffer. Both supernatant and sediment samples were detected using SDS-PAGE electrophoresis.

2.4 Purification of MBP-fused protein by affinity column chromatography

Protein fused with MBP was purified though amylose resin by affinity column chromatography. For ensuring the acceptable solubility of MBP-NIb, 500 mL culture was incubated at 20℃ with the induction of 0.4 mmol/L IPTG. All culture was centrifuged and treated by ultrasonic processing as shown in 2.3. With a 0.45 μm water-based membrane filter, the supernatant was filtered into a 50-mL microcentrifuge tube, which was placed on ice waiting for affinity column chromatography.

Affinity column chromatography was performed at 4℃. 2 mL amylose resin (NEB E8021S) was added into the column with the package buffer running out. Then at least 50 mL protein buffer was used to balance the column. Protein samples were added into the balanced column and went through the column through gravity action. After loading of all proteins, at least 50 mL protein buffer was used to wash the column. Finally, 10 mL elution buffer (20 mmol/L Tris-HCl (pH 7.4), 0.2 mol/L NaCl, 1 mmol/L EDTA, 10 mmol/L maltose) was used to elute the MBP-fused protein. All eluted products were collected using numbered tubes followed by SDS-PAGE electrophoresis to detect the quantity and quality of the MBP-NIb.

2.5 In vitro transcription with T7 RNA polymerase

With the plasmid as template, PCR was performed to amplify different PCR fragments corresponding 5′ terminal regions, internal regions and 3′ terminal regions of RNA1 and RNA2 of WYMV HC isolate. Purified PCR products were used to do in vitro transcription with T7 RNA polymerase (Promega). The 100 L reaction consists of DNA template 2 μg, 100 mmol/L DTT 10 μL, 20 mmol/L ATP 2.5 μL, 20 mmol/L GTP 2.5 μL, 20 mmol/L CTP 2.5 μL, 20 mmol/L UTP 2.5 μL, T7 RNA polymerase 5 μL, 2× T7 buffer 50 μL and ddH_2O. In vitro transcription was performed at 37℃ for 2 h. The target RNAs were separated through 1.5% agarose gel electrophoresis and purified by glass wool.

2.6 In vitro replication reaction

The 50 L reaction systems consist of 1 μg of RNA templates, 1 mol/L Tris-Cl (pH 8.2) 5 μL, 1 mol/L mol/$MgCl_2$ 0.5μL, 1 mol/L DTT 0.5 μL, 1 mol/L KCl 5 μL, 10 mg/mL yeast tRNA 1.75 μL, 20 mmol/L ATP 2.5 μL, 20 mmol/L GTP 2.5 μL, 20 mmol/L CTP 2.5 μL, 1 mmol/L UTP 0.5 μL, α-32P-UTP 0.5 μL, 12.5 μg of tested protein and ddH_2O. In vitro replication reaction was performed at 20℃ for 1.5 h. After the reaction, 70 μL ddH_2O and 120 μL of phenol/chloroform were added followed by centrifugation at 13000 r/min. The supernatant was precipitated by adding 2.4 times volume of NH_4Ac/isopropanol (5 mol/L NH_4Ac: isopropanol = 1∶5) and placed at -20℃ for at least 2 h. The precipitation was resuspended using 10 μL RNA loading buffer and separated in 5% PAGE gel containing 8 mol/L urea with 1500 mA for 1.5 h. Then, the gel was dried and exposed to a phosphorimager screen, followed by detection with the Typhoon FLA-7000(GE Healthcare).

3 Results

3.1 Construction of prokaryotic expression vector pMAL-WY-NIb

With the plasmid pUWR1-1 as template, 1.6 kb PCR products were amplified using primers WYM-R1-4921-BamHI-F and WYM-R1-6504-SalI-R (Fig. 1-A). The size of this PCR product is consistent with the length of the coding sequence of WYMV NIb (1584 nt). Both the purified PCR fragments and pMAL-C2X were digested with *Bam*H I and *Sal* I and ligated. After ligation, transformation and colony PCR, positive plasmid was digested to confirm the insertion of NIb coding sequences into pMAL-C2X. After digestion with *Bam*H I and *Sal* I, plasmid produced 6.0 kb and 1.6 kb of fragments (Fig. 1-B), which correspond with basic vector and NIb coding fragment, respectively. The positive plasmid pMAL-WY-NIb was also confirmed by DNA sequencing.

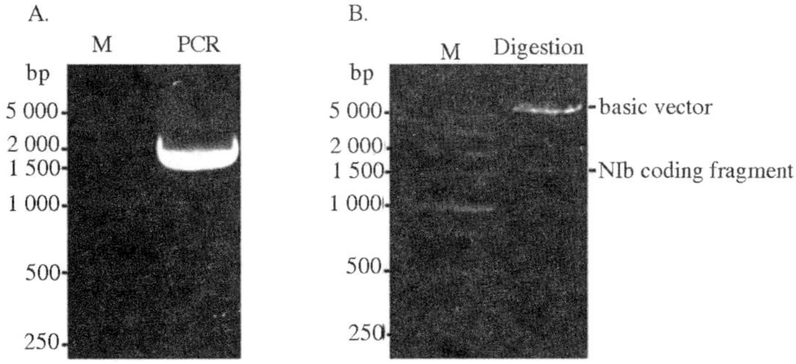

Fig. 1　Construction of prokaryotic expression plasmid containing WYMV NIb

A: PCR of WYMV NIb　B: Digestion of prokaryotic expression plasmid containing WYMV NIb

3.2　Optimization of prokaryotic expression of MBP-NIb fusion protein

Compared with the negative control, different concentration (0.2 to 1.0 mmol/L) of IPTG can induce the expression of MBP-NIb fused protein with the molecular weight of about 100 ku (Fig.2), which is consistent with the sum of the molecular weight of MBP (40 ku) and NIb (58 ku). Moreover, it is suggested that 0.4 mmol/L IPTG is optimal to induce the expression of MBP-NIb at 37℃ (Fig.2).

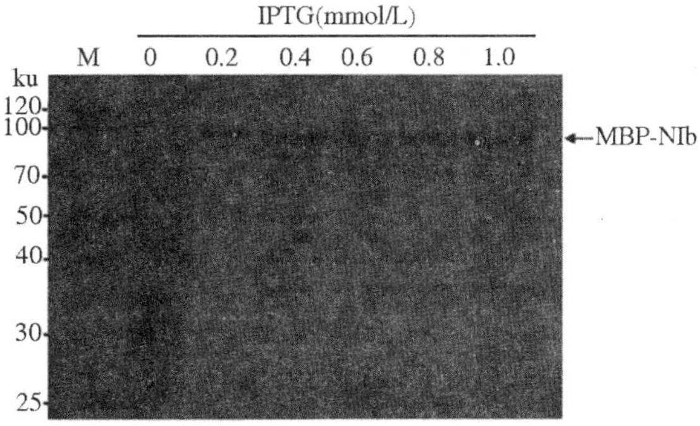

Fig. 2　Expression of NIb protein in *E. coli* Rosetta with the induction of IPTG.

M: Protein marker.

Solubility analysis was also performed to identify the optimal temperature of prokaryotic expression, which could ensure the acceptable solubility of MBP-NIb to meet the requirement of affinity chromatography. At 37℃, the ratio of soluble MBP-NIb is less than 1% (Fig. 4). It is suggested that MBP-NIb expressed at 37℃ prefer to form inclusion bodies, showing high insolubility (Fig. 3). At 20℃, the solubility of MBP-NIb was remarkably increased. The ratio of soluble MBP-NIb is about 30% (Fig. 3), which could meet the requirement of affinity chromatography. In addition, it can as far as possible ensure the activity of purified proteins due to the avoidance of extra process of the denaturation.

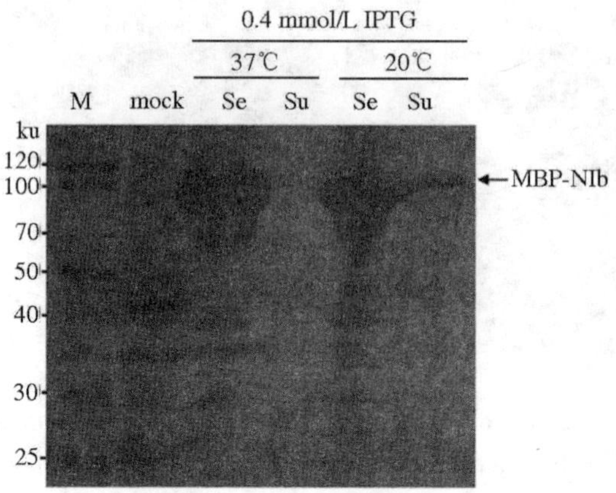

Fig. 3　Solubility analysis of MBP-NIb protein

M: Protein marker, mock: without the induction of IPTG, Se: Sediment, Su: Supernatant.

3.3　WYMV NIb-mediated in vitro replication system

To identify the activity and specificity of purified MBP-NIb through affinity chromatography, *in vitro* replication reaction was performed using 6 different RNA fragments as template. MBP-NIb can specifically recognized the 3′ fragment of WYMV RNA1 and RNA2, subsequently catalyzed the corresponding transcription of complementary strands, while MBP-NIb did not catalyze the in vitro replication of 5′ fragment and internal fragment of WYMV RNA1 and RNA2 (Fig.4). In addition, MBP did not catalyze the in vitro replication of the 3′ fragment of WYMV RNA1 and RNA2 (Fig.4). The activity of other proteins catalyzing *in vitro* replication of the 3′ fragment of WYMV RNA1 and RNA2 was also analyzed (Fig. 5). MBP-TB-RdRp containing RdRp of *Tobacco bushy top virus* (TBTV), which can catalyze *in vitro* replication of TBTV 3′ fragment (Lu et al., 2017), can't perform the in vitro replication of the 3′ fragment of WYMV RNA1 or RNA2 (Fig. 5). VPg, 14K and P3 of WYMV also can't perform the in vitro replication of the 3′ fragment of WYMV RNA1 or RNA2 (Fig.5). It is suggested that NIb of WYMV specifically preformed in vitro replication of the 3′ fragment of WYMV RNA1 and RNA2.

4　Discussion

The core component for setting up in vitro replication system is the replicase, which can be obtained from viral-infected cells or heterologous expression systems (Hong et al., 1996; Zhong et al., 2000; Panavas et al., 2002; Rajendran et al., 2002). Replicase purified from viral-infected cells possess the characteristic of posttranslational protein-modification such as phosphorylation, glycosylation and ubiquitination, while replicase proteins purified from heterologous expression systems such as *E. coli* easily possess high purity. In this study, we obtained the RdRp (NIb protein) from *E. coli* through prokaryotic expres-

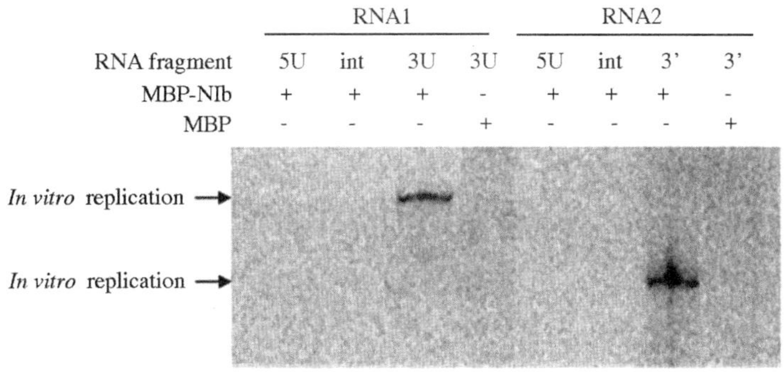

Fig. 4 In vitro replication mediated by NIb of WYMV.
RNA1 5U is 160 nt, RNA1 int is 160 nt, RNA1 3U is 260 nt;
RNA2 5U is 170 nt, RNA2 int is 160 nt, RNA2 3′ is 190 nt.

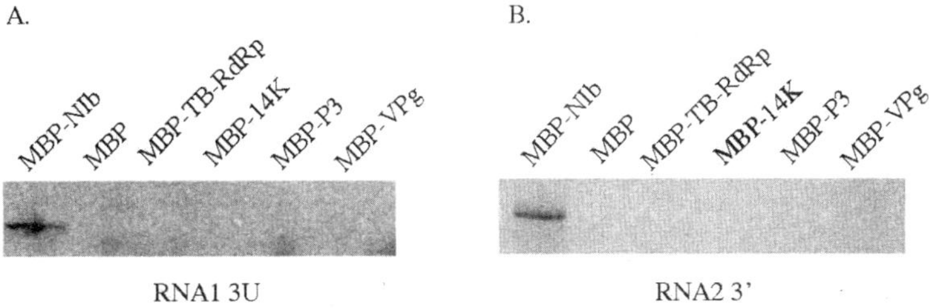

Fig. 5 Activity of different proteins on WYMV RNA1 and RNA2 in in vitro replication system.
A: Activity of different proteins on WYMV RNA1 3′ fragment. B: Activity of different proteins on WYMV RNA2 3′ fragment. MBP, maltose binding protein; NIb, NIb protein of WYMV; TB-RdRp, TBTV RNA-dependent RNA polymerase; 14K, 14K protein of WYMV; P3, P3 protein of WYMV, VPg, VPg protein of WYMV.

sion. The whole process is easily handled and repeated, meanwhile the purified RdRp presented the catalytic activity and RNA specificity (Fig.4 and Fig.5). The induction with IPTG under low-temperature is the key step to ensure the activity of purified proteins due to the avoidance of extra process of the denaturation.

Soil-borne wheat mosaic disease is one of the most economically important diseases on wheat in East Asia, especially in China. It was first reported in 1970s, and had now occurred largely and expanded gradually in Shandong, Zhejiang, Jiangsu, Henan, Hubei, Shanxi and other provinces(Li et al., 1997; Han et al., 2000; Sun et al., 2012). The management on soil-borne wheat mosaic disease is mainly depended on varietal resistance of wheat and transgenic wheat. The detailed mechanism how WYMV perform the genome replication is unclear. In this study, purified MBP-NIb presented higher specificity on 3′ fragment than other fragments of WYMV RNA1 and RNA2 (Fig.4 and Fig.5), which is also shown in research of other viruses (Hong et al., 1996; Panavas et al., 2002; Rajendran et al., 2002; Yuan et al., 2010; Lu et al., 2017). Using this *in vitro* replication system, further research will be performed on the core elements involved in replication of WYMV. Although VPg, 14K and P3 of other viruses are associated with *in vivo* accumulation, WYMV VPg, 14K or P3 can't solely catalyze *in vitro* replication of viral genome (Fig. 5). It suggests that these three proteins may affect *in vivo* viral accumulation through other processes such as translation, nucleocytoplasmic shuttling or production of replication-associated inclusion bodies (Sun et al., 2013; Sun et al., 2014; Li and Shirako, 2015).

Acknowledgments

We are grateful to Prof. Chenggui Han in China Agricultural University for providing the plasmid pUWR1-1. This work was supported by National Natural Science Foundation of China (31370179 and 31670147), Special Fund for Agro-scientific Research in the Public Interest in China (201303021) and Funds of Shandong "Double Tops" Program.

Compliance with ethics guidelines

The authors declare that they have no conflict of interest.

References

[1] Abe H, Tamada T. Parasite of polymyxa betae. Annals of Phytopathological Society of Japan, 1986, 52: 235-238.

[2] Buck K W. Comparison of the replication of positive-stranded RNA viruses of plants and animals. Adv. Virus Res., 1996, 47: 159-251.

[3] Chen J P. Research status and prospect of cereal viruses transmitted by Polymyxa graminis in China. Proc. Natural. Sci., 2005, 15: 524-533.

[4] Dong J H. Development of cDNA clones for RNAs in vitro transcription of *Wheat yellow mosaic virus* and their infections of cell culture systems. PhD thesis. Beijing: China Agricultural University, 2004.

[5] Ferrer-Orta C, Arias A, Escarmis C, Verdaguer N. A comparison of viral RNA-dependent RNA polymerases. Current Opinion in Structural Biology, 2006, 16: 27-34.

[6] Han C, Li D, Xing Y, *et al*. Wheat yellow mosaic virus widely occurring in wheat (*Triticum aestivum*) in China. Plant Dis., 2000, 84: 627-630.

[7] Hodge K, Tunghirun C, Kamkaew M, *et al*. Identification of a conserved RNA-dependent RNA polymerase (RdRp)-RNA interface required for Flaviviral replication. J. Biol. Chem., 2016, 291(33): 17437-17449.

[8] Hong Y, Hunt A G. RNA polymerase activity catalyzed by a potyvirus-encoded RNA-dependent RNA polymerase. Virology, 1996, 226:146-151.

[9] Inouye T. Filamentous particle as the causal agent of yellow mosaic disease of wheat. Nogaku Kenkyu, 1969, 53: 61-68.

[10] Klein P G, Klein R R, Rodriguez-Cerezo E, *et al*. Mutational analysis of the tobacco vein mottling virus genome. Virology, 1994, 204(2): 759-769.

[11] Koonin E V, Gorbalenya A E, Chumakov K M. Tentative identification of RNA-de-

pendent RNA polymerases of dsRNA viruses and their relationship to positive strand RNA viral polymerases. FEBS Lett., 1989, 252 (1-2): 42-46.

[12] Kuhne T. Soil-borne viruses affecting cereals-known for long but still a threat. Virus Res., 2009, 141: 174-183.

[13] Lai M M. Cellular factors in the transcription and replication of viral RNA genomes: a parallel to DNA-dependent RNA transcription. Virology, 1998, 244: 1-12.

[14] Langenberg W G, Zhang L. Immunocytology shows the presence of Tobacco etch virus P3 protein in nuclear inclusions. J. Struct. Biol., 1997, 118: 243-247.

[15] Lewandowski D J, Dawson W O. Functions of the 126- and 183-ku proteins of tobacco mosaic virus. Virology, 2000, 271: 90-98.

[16] Li D, Han C, Xing Y, *et al*. Identification of the wheat yellow mosaic virus occurring in China by RT-PCR. Acta Phytopathol Sin., 1997, 27: 303-307.

[17] Li H, Shirako Y. Association of VPg and eIF4E in the host tropism at the cellular level of *Barley yellow mosaic virus* and *Wheat yellow mosaic virus* in the genus Bymovirus. Virology, 2015, 476: 159-167.

[18] Liu W W, Zhao X J, Zhang P, et al. A one step real-time RT-PCR assay for the quantitation of Wheat yellow mosaic virus (WYMV). Virol. J., 2013, 10: 173.

[19] Lu X, Yu C, Wang G, et al. *In vitro* replication system mediated by RNA-dependent RNA polymerase (RdRp) in *Tobacco bushy top virus* (in Chinese). Acta Phytopathol. Sin., 2017, doi:10.13926/j.cnki.apps.000116

[20] Nagy P D. Viral sensing of the subcellular environment regulates the assembly of new viral replicase complexes during the course of infection. J. Virol., 2015, 89(10): 5196-5199.

[21] Ohki T, Netsu O, Kojima H, *et al*. Biological and genetic diversity of *Wheat yellow mosaic virus* (Genus Bymovirus). Phytopathology, 2014, 104(3): 331-319

[22] Panavas T, Pogany J, Nagy P D. Analysis of minimal promoter sequences for plus-strand synthesis by the Cucumber necrosis virus RNA-dependent RNA polymerase. Virology, 2002, 296(2): 263-274.

[23] Rajendran K S, Pogany J, Nagy P D. Comparison of *Turnip crinkle virus* RNA-dependent RNA polymerase preparations expressed in *Escherichia coli* or derived from infected plants. J. Virol., 2002, 76(4): 1707-1717.

[24] Sawada E. Control of *wheat yellow mosaic virus*. Plant Prot., 1927, 14: 444-449.

[25] Schaad M C, Jensen P E, Carrington J C. Formation of plant RNA virus replication complexes on membranes: role of an endoplasmic reticulum-targeted viral protein. The EMBO J., 1997, 16(13): 4049-4059.

[26] Sun B, Yang J, Sun L, *et al*. Distribution and dynamics of *Polymyxa graminis* transmitted wheat viruses in China. J. Triticeae Crops, 2012, 31: 969-973.

[27] Sun L, Bian J, Andika I B, et al. Nucleo-cytoplasmic shuttling of VPg encoded by *Wheat yellow mosaic virus* requires association with the coat protein. J. Gen. Virol.,

2013, 94: 2790-2802.

[28] Sun L, Andika I B, Shen J, *et al*. The P2 of Wheat yellow mosaic virus rearranges the endoplasmic reticulum and recruits other viral proteins into replication-associated inclusion bodies. Molecular plant pathology, 2014, 15(5): 466-478.

[29] Tao J, Qin J, Xiao J, *et al*. Studies on the soil-borne yellow mosaic virus of wheat in Sichuan (in Chinese). Acta Phytopathol. Sin., 1980, 10(1): 15-27.

[30] te Velthuis A J. Common and unique features of viral RNA-dependent polymerases. Cellular and Molecular Life Sciences, 2014, 71(22): 4403-4420.

[31] Wang N, Liu G, Lu X. Preliminary study on Wheat spindle streak mosaic virus place in China (in Chinese). Sichuan Agricultural Science and Technology, 1980,1: 34-35

[32] Yu S, Chen Z, Xu L. Wheat yellow mosaic virus disease in China (in Chinese). Acta Phytophylacica Sinica, 1986, 13(4): 217-219.

[33] Yuan X, Shi K, Young M Y, *et al*. The terminal loop of a 3′ proximal hairpin plays a critical role in replication and the structure of the 3′ region of *Turnip crinkle virus*. Virology, 2010, 402(2): 271-280.

[34] Zhong W, Uss A S, Ferrari E, *et al*. *De novo* initiation of RNA synthesis by hepatitis C virus nonstructural protein 5B polymerase. J. Virol., 2000, 74: 2017-2022.

[35] Zhang Z Y, Liu X J, Li D W, *et al*. Rapid detection of wheat yellow mosaic virus by reverse transcription loop mediated isothermal amplification. Virol. J., 2011, 8: 550.

两种类型的黄瓜花叶病毒卫星 RNA 山东分离物的序列和结构分析

刘珊珊[1]，曹欣然[1,2]，于成明[1]，耿国伟[1]，李向东[1]，原雪峰[1*]

([1]山东农业大学植物保护学院植物病理系，山东省农业微生物重点实验室，泰安 271018；
[2]烟台市农业技术推广中心，烟台 264001)

摘要：黄瓜花叶病毒（*Cucumber mosaic virus*，CMV）卫星 RNA（satellite RNA，satRNA）的存在通常影响 CMV 的致病性。本研究设计了检测 satCMV 的简并引物，通过 RT-PCR 从山东省泰安地区的 3 种不同自然发病寄主中检测到 satCMV 的存在，发现存在两类长度有明显差异的 satCMV：白菜与萝卜中克隆的 satCMV 为 339 个核苷酸，二者的核苷酸一致性为 99.41%，而烟草中克隆的 satCMV 为 383 个核苷酸，与来自白菜和萝卜的 satCMV 的核苷酸序列一致性均为 71.06%。系统进化分析表明这两类长度有明显差异的 satCMV 亲缘关系较远。利用 mFold 软件分析了 2 类不同长度的 CMV 卫星 RNA 基因组及其互补链的结构特征，暗示了 2 类 satCMV 与 CMV 互作的活跃度差异以及核心结构域的差别。两类 satCMV 全基因组的克隆以及结构分析为研究不同类型 satCMV 与 CMV 的互作及其对 CMV 致病性的影响奠定了基础。

关键词：黄瓜花叶病毒；卫星 RNA；系统进化；RNA 结构

Analysis on the sequence and structure of two types of satellite RNA *Cucumber mosaic virus* from Shandong province

Shanshan Liu[1], Xinran Cao[1,2], Chengming Yu[1], Guowei Geng[1], Xiangdong Li[1], Xuefeng Yuan[1*]

([1] *Department of Plant Pathology, College of Plant Protection, Shandong Agricultural University; Shandong Province Key Laboratory of Agricultural Microbiology, Tai'an 271018, China;*
[2] *Yantai Agricultural Technology Extension Center, Yantai 264001, China*)

Abstract: Satellite RNA interfers the virulence of *Cucumber mosaic virus* (CMV) in

---

基金项目：国家自然科学基金（31670147，31370179）；山东省“双一流”奖补资金资助经费（SYL2017XTTD11）

通讯作者：原雪峰，教授，主要从事分子植物病毒学研究；E-mail：snowpeak77@163.com

第一作者：刘珊珊，女，山东泰安人，硕士研究生，从事植物病毒学研究；E-mail：shansd1218@126.com。

different ways. In the study, satCMV was detected using selfdesigned primers from three hosts in Tai'an city of Shandong province. These three new satCMV isolates have two types of genome length as 339 nt and 383 nt. Sequence identity between TA-ca and TA-ra with 339 nt genome is 99.41%, and sequence identities between these two isolates and TA-tb with 383 nt genome are both 71.06%. Phylogenetic analysis suggests the evolution differentiation for these two types of satCMV with different length of genome. Meanwhile, RNA structures of these two type satCMV and its complementary strand were predicted and it is suggested the potential difference of liveness and core-motif between these two types of satCMV. Genome cloning and structure analysis on these two types of satCMV provide the material and data for future research on molecular cross-talk between satCMV and CMV.

Keywords: *Cucumber mosaic virus*; satellite RNA; phylogenetic tree; RNA structure

黄瓜花叶病毒(*Cucumber mosaic virus*, CMV)由 Doolittle 和 Jagger 于 1916 年首次报道[1,2],能侵染 100 多个科的 2000 多种植物,包括单子叶植物和双子叶植物。由 CMV 感染引起的宿主物种症状包括花叶,褪绿,矮化,畸形和全身性坏死,早期发病的植株严重矮化,生长缓慢,不能正常开花结实。CMV 在世界范围内分布广泛,特别是在温带到热带气候区;该病毒可以经过多达 80 种蚜虫传播,还能经过种子传播,是寄主最多、分布最广、最具经济危害的植物病毒之一[3]。

CMV 属于雀麦花叶病毒科(*Bromoviridae*)黄瓜花叶病毒属(*Cucumovirus*),是典型的单链正义 RNA(+ ssRNA)病毒,基因组由 RNA1、RNA2 和 RNA3 组成,共编码 5 个蛋白。RNA1 编码 1a 蛋白,是病毒基因组复制所必需的蛋白。RNA2 编码 2a 与 2b 蛋白[4],2a 蛋白具有 RNA 依赖的 RNA 聚合酶(RdRp)活性,与 1a 蛋白共同参与病毒的体内复制过程,影响寄主的症状[5];2b 蛋白由亚基因组 RNA4A 产生,2b 蛋白对寄主植物转录后基因沉默具有抑制作用,是一种基因沉默抑制因子,能够最大限度地保护病毒基因组不被寄主植物降解[6]。RNA3 编码 3a 蛋白即运动蛋白 MP 和外壳蛋白 CP,MP 与病毒在寄主细胞间移动和长距离移动有关[7],CP 由亚基因组 RNA4 产生,与病毒颗粒组装、蚜虫传播、病毒血清型及寄主症状表现有关[4,8-11]。

对于某些 CMV 株系,还含有卫星 RNA(satellite RNA, satRNA),是一类非编码单链 RNA 分子,长度有较大差异,大小从 246 nt (KT382260) 到 405 nt (D28559)不等,是病毒基因组以外的遗传因子,必须依赖于辅助病毒 CMV 的 RdRp 完成基因组复制,能够和 CMV 一起借助蚜虫等媒介进行传播[7]。现已报道影响植物症状的 satCMV 有 3 种,包括 D-satRNA、Y-satRNA 及 SD-satRNA。D 类 satRNA 能够影响植物维管细胞发育并且能导致程序性细胞死亡[12]。Y 类 satRNA 能够引起许多茄科植物(如本氏烟、烟草、甜椒)的黄化,当病毒 Y-satRNA 侵染烟草产生小干扰 RNA(siRNA),由于病毒 siRNA 与叶绿素合成关键酶 CHlI 的 mRNA 存在 22 nt 的同源互补区域,因而随着 satRNA 的 siRNA 产生,CHlI 的 mRNA 会被降解,从而影响了叶绿素合成并导致烟草叶片的黄化症状[13,14]。SD 类 satRNA 可能会导致 CMV 症状的衰减[15],在感染 SD 类 satRNA 的本氏烟组织中,RNA4A 和其编码的 2b 蛋白的积累量减少[16],可能是 satCMV 产生的 siRNA 可以靶向 RNA4A 区域。也有报道提到了 satCMV 与 CMV 竞争 RdRp 从而造成 CMV 致病力下降[17],然而关于不同类型 satCMV 与 CMV RdRp 的互作研究尚未明确。

本研究以采自山东泰安的 3 种感病寄主为材料,在检测 CMV 存在的基础上,设计简并引物检测 satCMV，发现了 2 种长度明显差异的 satCMV，对这 2 种不同类型的 satCMV 进行了全基因组克隆以及序列、结构和进化分析，为研究 satCMV 与 CMV 的互作奠定基础。

1 材料与方法

1.1 感病材料、载体和菌种

表现花叶症状的白菜、萝卜、烟草样品，于 2016 年春采自山东省泰安市泰山区。大肠杆菌 DH5α 菌株由本实验室保存。pMD18-T 载体购自 TaKaRa 公司。

1.2 CMV 和 satCMV 的检测

采用 TransZol 试剂（TransGen Biotech）分别提取表现花叶症状的白菜、萝卜、烟草的植物总 RNA，分别针对 CMV 和 satCMV 进行 RT-PCR 检测。RT 的反应条件为 42℃，1.5 h；PCR 的反应条件为 94℃，5 min；94℃，40 s，51℃，40 s，72℃，30 s，30 个循环；72℃，10 min。引物 CM-R3-119-F 和 CM-R3-768-R 用于 CMV 的检测，引物 satCMV-int-5F 和 satCMV-int-3R 用于 satCMV 的检测；所用引物的具体信息见表 1。

Table 1 Primers used in this study

Primers	Sequences(5′→3′)
CM-R3-119-F	AGGCATGGCTTTCCAAGGTAC
CM-R3-768-R	CTGACGGTTTCGTTTGCTCAGC
satCMV-int-5F	GCGGTGTGGGWTAMCTCC，W=A/T，M=C/A
satCMV-int-3R	GCYTCTCCCTGYGTGCG，Y=T/C
CMV-satRNA-F	GTTTTGTTTGTTGGAGAATTGCGRGAGGGGTTAT，R=A/G
CMV-satRNA-R	GGGTCCTGTAGAGGAATGWTAGACATTCACGGA，W=A/T

1.3 satCMV 的全基因组克隆

利用白菜、萝卜、烟草病样的植物总 RNA 为模板，进行 RT-PCR。RT 反应条件为 42℃，1.5 h，所用引物为 CMV-satRNA-R；PCR 条件为 94℃，40 s，56℃，40 s，72℃，30 s，30 个循环；72℃，10 min。所用简并引物为 CMV-satRNA-F 和 CMV-satRNA-R（表 1）。所得 PCR 产物经胶回收与 pMD18-T 载体连接，阳性克隆利用通用引物 M13F 测序。

1.4 序列比对、序列一致性和系统进化树分析

测序所得序列利用 NCBI 进行 BLAST 比对；采用 DNAMAN 软件分析核苷酸序列一致性；采用 MEGA5.0[18] 软件的邻接法（neighbor-joining，NJ）进行系统进化树构建，所有设置依照默认值。

1.5 RNA 结构预测

采用在线软件 mFold 2.3 版（http://unafold.rna.albany.edu/? q=mfold/RNA-Folding-Form 2.3）分析 RNA 二级结构，温度设定为 25℃，其他设置依照默认值。

2 结果与分析

2.1 三种寄主中 CMV 和 satCMV 的检测

对于表现花叶症状的白菜、萝卜、烟草样品，首先检测 CMV 是否存在。利用特异性引物 CM-R3-119-F 和 CM-R3-769-R 检测 CMV RNA3 的部分片段，RT-PCR 结果显示扩增的特

异性条带大小约为 650 bp(图 1-A),克隆后测序结果的 BLAST 比对也证实 PCR 产物为 CMV RNA3 的相应片段,并提交 NCBI,登录号分别是 KY646100, KY646102 和 KY646103。

在明确这 3 种表现花叶症状的寄主中均存在 CMV 的前提下,进行 satCMV 的检测。首先下载 NCBI 中已报道的 173 个 satCMV 序列(具体信息略),利用 DNAMAN 软件进行序列比对来定位 satCMV 的序列保守区段,然后根据保守区段设计简并引物 satCMV-int-5F 和 satCMV-int-3R(表 1);根据已报道序列的信息,此对引物的扩增片段大小为 220～280 bp。利用该对引物进行 RT-PCR 检测,从 3 种不同寄主中扩增得到长度有差异的 250 bp 左右的片段(图 1-B),测序结果和 BLAST 比对表明,从 3 种不同寄主中所扩增的 250 bp 左右的序列与 satCMV 的同源性最高,说明白菜、萝卜、烟草的感病样品中均存在 satCMV,但长度有差异。

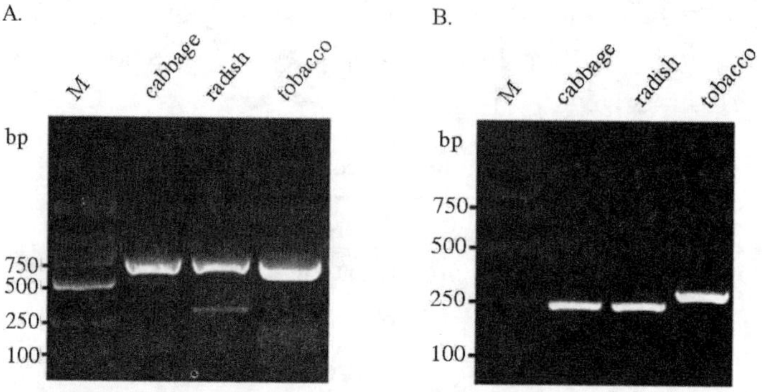

Fig. 1　Detection of CMV and satCMV in different hosts from Tai'an city of Shandong province

A: RT-PCR of CMV RNA3 partial fragment. M: DL2504-100 DNA ladder;

B: RT-PCR of satCMV partial fragment. M: DL5000 DNA DNA ladder.

2.2　三种寄主中 satCMV 的全基因组序列克隆

利用对应于 satCMV 的 5′和 3′末端序列的简并性引物进行 satCMV 基因组的 RT-PCR,结果显示 3 种寄主中的 satCMV 的全长也存在长度差异。白菜 satCMV 与萝卜 satCMV 的全长约为 350 bp,烟草 satCMV 的全长约为 400 bp(图 2)。PCR 产物克隆到 pMD18-T 载体中,利用通用引物 M13F 的测序结果显示白菜中 satCMV 与萝卜中 satCMV 全长均为 339 个核苷酸,烟草中 satCMV 全长为 383 个核苷酸。提交 GenBank,登录号分别为白菜分离物(TA-ca,MF142363)、萝卜分离物(TA-ra,MF142364)和烟草分离物(TA-tb,MF142365)。

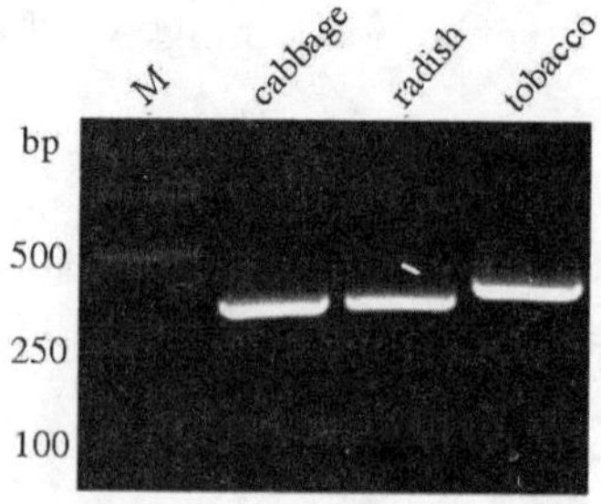

Fig. 2　RT-PCR of satCMV genome in different hosts from Tai'an city of Shandong province

M: DL2504-100 DNA ladder

2.3 satCMV 序列、系统进化和 RNA 结构分析

利用 DNAMAN 软件对 3 种寄主中 satCMV 进行序列比对，结果显示 satCMV TA-ca 和 satCMV TA-ra 的序列一致率为 99.41%，而 satCMV TA-tb 与前二者的序列一致率均为 71.06%。其中序列差异区域主要位于 satCMV 基因组的中间区域（图 3）。

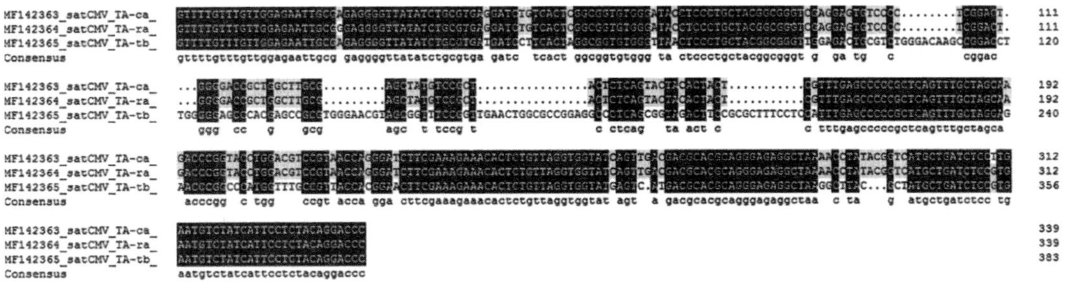

Fig. 3 Sequence alignment of three new satCMV isolates in Tai'an city of Shandong province

为分析新克隆 satCMV 的进化关系，选取 51 条已报道的 satCMV 基因组序列（表 2），基于 satCMV 基因组序列构建系统进化树。系统进化树分为 2 个大组，第 1 大组分为 3 个亚组（图 4）。来自山东泰安的 TA-ca（白菜分离物）与 TA-ra（萝卜分离物）聚为单独一支，属于第一大组第一亚组，与包括塞尔维亚的 KM358139、以色列的 U43889、中国的 AY589729 和 HQ283395 的另一支的亲缘关系最近（图 4）。来自山东泰安的 TA-tb（烟草分离物）则成为单独一支，属于第二大组，与包括中国的 AF451897、DQ412733 和 EF363687 的另一支的亲缘关系最近（图 4）。

Table 2 Summary of satCMV isolates for construction of phylogenetic tree

Accession Number	Length (nt)	Host	Country	Accession Number	Length (nt)	Host	Country
AB072504	383	Tomato	Japan	JN029953	339	Snap bean	USA
AB448703	334	*Solanum tuberosum*	Syria	KC415034	338	*Phaseolus vulgaris*	USA
AB570294	398	*Capsicum annuum*	South Korea	KC985209	339	*Pratia pedunculata*	USA
AB570295	386	*Physalis alkekengi*	South Korea	KC985210	339	*Pratia pedunculata*	Mexico
AF451896	368	*Raphanus sativus*	China	KJ127541	341	*Nicotiana glauca*	Mexico
AF451897	384	*Raphanus sativus*	China	KM358139	320	Tomato	Serbia
AJ457164	394	Pea	China	KP719209	317	Tomato	Serbia
AY589729	336	*Capsicum annuum*	China	KT382259	282	Tomato	Greece
D00541	338	Tomato Tobacco	England	KT382261	281	*Phaseolus vulgaris* L.	Greece
D00699	338	*Petasites japonicum* Miq	Japan	KT382262	298	*Vigna unguiculata* L.	Greece
D10037	335	Tobacco Tomato	Japan	KT382264	284	*Lens culinaris* Medik	Greece
D10039	384	Tobacco	Japan	KT382266	299	*Phaseolus coccineus* L.	Greece
D11053	368	Tomato	Japan	KU144672	341	*Dianthus caryophyllus*	Mexico
D28559	405	Tomato	Japan	KU169300	349	*Physalis ixocarpa*	Mexico

Continue Table 2

Accession Number	Length (nt)	Host	Country	Accession Number	Length (nt)	Host	Country
DQ070747	336	*Beta vulgaris*	China	U31660	387	Tomato	USA
DQ249297	383	Tomato	China	U43889	339	*Musa acuminata*	—
DQ412733	387	Tobacco	China	X51465	340	Tobacco	USA
DQ975201	336	Tobacco	USA	X54065	337	Tobacco	Germany
EF363687	385	Tomato	China	X65455	390	Tomato	Italy.
EF363688	384	pepper	China	X86421	334	Tobacco	Italy
EU022373	335	Tobacco	China	X86423	334	Tomato	Italy
HQ283395	336	Tomato	China	X86424	338	*Nicotiana glauca*	Italy
J02060	336	*Cucumis sativus*	Australia	X86425	336	Celery	Italy
J02061	334	Tomato	USA	X89004	334	Tomato	Croatia
JF918971	339	*Vinca minor*	USA	Z75871	339	Tomato	Spain
JF918972	389	*Vinca minor*	USA				

satCMV 是依赖于 CMV 的 RdRp 来完成基因组的复制，RdRp 与 satCMV 的识别取决于 satCMV 的 RNA 结构特征。3 种寄主中的 satCMV 有 2 种长度类型，此长度差异是否造成 satCMV 的 RNA 结构剧烈变化？利用 MFold 在线软件预测了白菜和烟草中 satCMV 基因组及其互补链的 RNA 结构。对 satCMV TA-ca 分离物和 TA-tb 分离物的基因组及其互补链的结构预测表明，均存在 3 种类型的结构：三段式（整体结构由 5′端的结构、3′端的结构以及中间区域结构组成），两段式（整体结构由 5′端的结构和 3′端的结构）和 5′端-3′端偶联型结构（图 5-A 和图 5-B）。对于 satCMV TA-ca 分离物基因组而言，25 个预测结构中包括 12 个三段式结构(48.0%)、3 个两段式结构(12.0%)、10 个 5′端-3′端偶联型结构(40.0%)；对于 satCMV TA-tb 分离物基因组而言，30 个预测结构中包括 8 个三段式结构(26.7%)、16 个两段式结构(53.3%)、6 个 5′端-3′ 端偶联型结构(20.0%)（图 5-A，表 3）。卫星 RNA 的基因组复制依赖辅助病毒的 RdRp 完成，主要通过 RdRp 与基因组模版的 3′末端特定结构的结合实现；而 5′端-3′端偶联型结构是一种结构比较封闭的结构，是不适合 RdRp 与 3′末端特定结构的结合的。satCMV TA-ca 的 5′端-3′端偶联型结构的比例(40.0%)高于 satCMV TA-tb 的 5′端-3′端偶联型结构的比例(20.0%)，暗示了 satCMV TA-ca 相对 satCMV TA-tb 而言，封闭型结构比例高。对于非封闭性（三段式结构和两段式结构）结构而言，satCMV TA-ca 和 satCMV TA-tb 中三段式结构和两段式结构的相对比例有所不同（表 3），但其 3′端的结构的核心结构（长茎环）相同（图 5-A），暗示 satCMV TA-ca 和 satCMV TA-tb 的非封闭性结构与 CMV RdRp 的互作的相似性，即以基因组为模板产生互补链的效率相似。

对于 satCMV TA-ca 分离物的基因组互补链而言，17 个预测结构中包括 2 个三段式结构(11.8%)、4 个两段式结构(23.5%)、11 个 5′端-3′端偶联型结构(64.7%)；对于 satCMV TA-tb 分离物基因组互补链而言，36 个预测结构中包括 10 个三段式结构(27.8%)、21 个两段式结构(58.3%)、5 个 5′端-3′端偶联型结构(13.9%)（图 5-B 和表 3）。satCMV TA-ca 基因组互补链的 5′端-3′端偶联型结构的比例(64.7%)高于 satCMV TA-tb 基因组互补链的 5′端-3′端

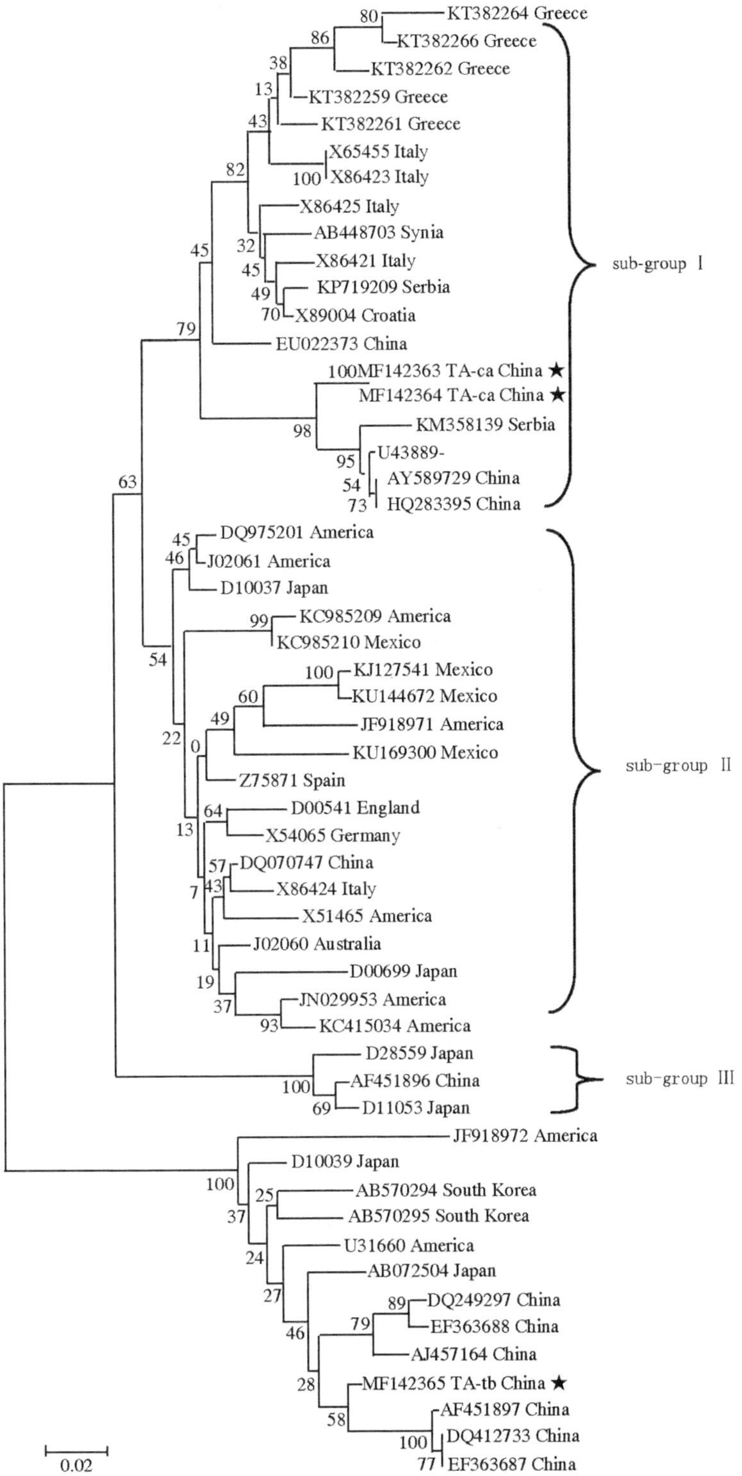

Fig. 4　Phylogenetic tree based on satCMV sequence

Note: Phylogenetic tree was constructed by the neighboring-joining (NJ) method with default setting. Black stars indicate three new satCMV isolates from Tai'an city in Shandong province.

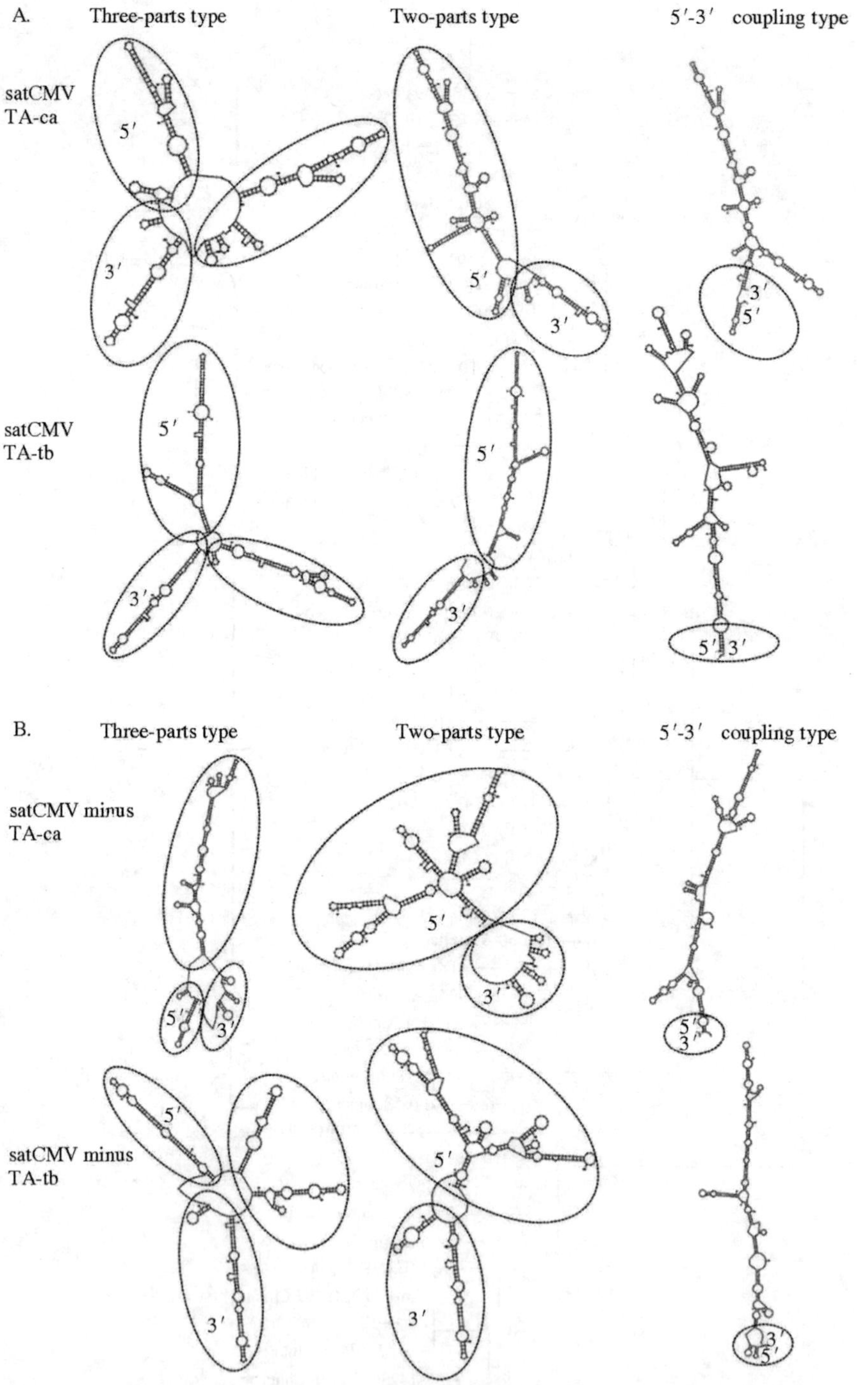

Fig. 5 RNA secondary structures of different satCMV and its complementary strand

A: RNA secondary structure of satCMV TA-ca isolate and satCMV TA-tb isolate;

B RNA secondary structure of their complementary strands of satCMV TA-ca isolate and satCMV TA-tb isolate.

Note: 5′ indicates 5′ terminal part; 3′ indicates 3′ terminal part;

minus indicates complementary strand of satCMV genome.

Table 3 Summary of RNA structures of two satCMV and its complementary strand

Constructs	Total numbers of predicted structures	Three-parts type	Two-parts type	5´-3´ coupling type
satCMV TA-ca (MF142363)	25	12 (48.0%)	3 (12.0%)	10 (40.0%)
satCMV TA-tb (MF142365)	30	8 (26.7%)	16 (53.3%)	6 (20.0%)
satCMV TA-ca minus	17	2 (11.8%)	4 (23.5%)	11 (64.7%)
satCMV TA-tb minus	36	10 (27.8%)	21 (58.3%)	5 (13.9%)

Note: numbers in bracket indicates the percentage in total predicted structures.

偶联型结构的比例(13.9%),暗示了 satCMV TA-ca 基因组互补链相对 satCMV TA-tb 基因组互补链而言,封闭型结构比例高。对于非封闭性(三段式结构和两段式结构)结构而言,satCMV TA-ca 基因组互补链和 satCMV TA-tb 基因组互补链中三段式结构和两段式结构的相对比例有所不同(表 3),并且其 3´端的结构的核心结构存在明显差异,satCMV TA-ca 基因组互补链 3´端的结构包括多个小茎环,而 satCMV TA-tb 基因组互补链 3´端的结构主要是一个大的长茎环(图 5-B);暗示 satCMV TA-ca 基因组互补链和 satCMV TA-tb 基因组互补链的非封闭性结构与 CMV RdRp 的互作存在明显差异,即以基因组互补链为模板新的基因组的效率差别较大。

3 讨论

不同类型卫星 RNA 影响 CMV 致病性的分子机制差别很大,主要包括 satCMV 与 CMV 竞争复制酶降低 CMV 的复制效率、satCMV 产生的 siRNA 可以靶向 CMV 的部分序列从而降解 CMV 序列、satCMV 产生的 siRNA 可以靶向植物内源序列影响植物的正常生长发育等[12-17]。然而,不同类型 satCMV 与 CMV 尤其是 RdRp 的分子互作机制尚未明确。

本研究根据已报道的 satCMV 序列设计了用于扩增 satCMV 部分序列的简并引物,并利用此简并引物从山东泰安的 3 种植物中克隆到长度有差异的两种类型 satCMV(图 2),序列分析和系统进化分析表明了 2 种 satCMV 的进化分化以及主要差异区(图 3 和图 4)。关于这 2 种类型 satCMV 及其互补链的 RNA 结构预测和分析表明:①两种类型 satCMV 的结构活跃度不同。长度较短的 TA-ca 相对于 TA-tb 而言,其 RNA 结构比较封闭,表现在 5´端-3´端偶联型结构的比例较高,无法形成 3´端独立的结构域被 CMV RdRp 识别。②两种 satCMV 的长度差异对于基因组和基因组互补链的影响不同。长度差异对基因组链的影响较小,表现在非封闭型结构中 3´端结构域的核心结构在两种 satCMV 中无明显变化(图 5-A);而对于基因组互补链的影响明显,表现在非封闭型结构中 3´端结构域的核心结构发生了显著变化(图 5-B)。③两种 satCMV 的长度差异可能影响 CMV RdRp 与其基因组互补链的识别效率,即影响由基因组互补链产生新的基因组的过程。对由基因组产生互补链的过程的影响较小。这两种类型 satCMV 的克隆和结构分析为将来研究 satCMV 与 CMV 尤其是 RdRp 的互作提供了材料和数据基础。

参考文献

[1] Doolittle S P. A new infectious mosaic disease of cucumber. Phytopathology, 1916, 6 (2): 145-147.

[2] Jagger I C. Experiments with the cucumber mosaic disease. Phytopathology, 1916, 6(2): 149-151.

[3] Xu Z, Barnett O W. Identification of a cucumber mosaic virus strain from naturally infected peanuts in China. Plant Disease, 1984, 68(5): 386-389.

[4] Palukaitis P, Roossinck M J, Dietzgen R G, et al. Cucumber mosaic virus. Advances in Virus Research, 1992, 41: 281-348.

[5] Hampton R O, Francki R I B. RNA-1 dependent seed transmissibility of cucumber mosaic virus in Phaseolus vulgaris. Phytopathology, 1992, 82(2): 127-130.

[6] Csorba T, Kontra L, Burgyán J. Viral silencing suppressors: tools forged to fine-tune host-pathogen coexistence. Virology, 2015, 479: 85-103.

[7] Palukaitis P and Garcia-Arenal F, Cucumoviruses. Advances in Virus Research, 2003, 62: 241-323.

[8] Edwardson J R, Christie R G. CRC handbook of viruses infecting legumes. CRC Press, 1991.

[9] Roossinck M J. Cucumber mosaic virus, a model for RNA virus evolution. Molecular Plant Pathology, 2001, 2(2): 59-63.

[10] Chen B, Francki R I B. Cucumovirus transmission by the aphid *Myzus persicae* is determined solely by the viral coat protein. Journal of General Virology, 1990, 71(4): 939-944.

[11] Ding S W, Anderson B J, Haase H R, *et al*. New overlapping gene encoded by the cucumber mosaic virus genome. Virology, 1994, 198(2): 593-601.

[12] Xu P, Roossinck M J. Cucumber mosaic virus D satellite RNA-induced programmed cell death in tomato. The Plant Cell, 2000, 12(7): 1079-1092.

[13] Shimura H, Pantaleo V, Ishihara T, et al. A viral satellite RNA induces yellow symptoms on tobacco by targeting a gene involved in chlorophyll biosynthesis using the RNA silencing machinery. PLoS Pathogens, 2011, 7(5): e1002021.

[14] Smith N A, Eamens A L, Wang M B. Viral small interfering RNAs target host genes to mediate disease symptoms in plants. PLoS Pathogens, 2011, 7(5): e1002022.

[15] Hou W N, Duan C G, Fang R X, *et al*. Satellite RNA reduces expression of the 2b suppressor protein resulting in the attenuation of symptoms caused by Cucumber mosaic virus infection. Molecular Plant Pathology, 2011, 12(6): 595-605.

[16] Roossinck M J, Sleat D, Palukaitis P. Satellite RNAs of plant viruses: structures and biological effects. Microbiological Reviews, 1992, 56(2): 265-279.

[17] Simon A E, Roossinck M J, Havelda Z. Plant virus satellite and defectiveinterfering RNAs: new paradigms for a new century. Annu. Rev. Phytopathol., 2004, 42: 415-437.

[18] Tamura K, Peterson D, Peterson N, *et al*. MEGA5: molecular evolutionary genetics analysis using maximum likelihood, evolutionary distance, and maximum parsimony methods. Molecular Biology and Evolution, 2011, 28(10): 2731-2739.

15 种植物提取物对苹果树腐烂病菌的抑制作用

杨树，薛应钰*

（甘肃农业大学植物保护学院，兰州 730070）

摘要：为探究不同植物提取物对苹果树腐烂病菌的抑制作用，本试验以 10 个科 15 种植物粗提物为研究材料，测定了其对苹果树腐烂病的抑菌活性，并挑取受抑制菌丝进行显微观察。结果表明：供试的 15 种植物的提取物对苹果树腐烂病菌均表现出了不同程度的抑制作用，其中甘草的乙醇提取物对苹果树腐烂病菌的抑制率最高，达 90% 以上；柴胡、合欢的乙醇提取物对供试的苹果树腐烂病菌的抑制率次之，均在 80% 以上。经显微镜下观察发现，利用甘草、柴胡、合欢提取物处理后的菌丝均呈现畸形。

关键词：植物提取物；苹果树腐烂病菌；抑菌活性

Inhibition effect of extracts of 15 kinds of plant against *Valsa mali*

Shu Yang, Yingyu Xue*

（*College of Plant Protection*, *Gansu Agricultural University*, *Lanzhou 730070*, *China*）

Abstract: To study the different plant extracts antifungal activity to *Valsa mali*, In this study, the crude extracts of 15 plants were used as the research materials, and the anti-bacterial activity against apple tree rot was determined. And select the inhibitory hyphae for microscopic observation. The results show that the extracts of 15 plants have all anti-fungal activity to *Valsa mali* in different degrees. The ethanol extract of licorice had the highest inhibitory rate to the rotting strain of *Valsa mali*, which was more than 90%. *Albizia julibrissin* and *Bupleurum chinensis*. enthanol extract of tested antifungal activity followed *Glycyrrhiza uralensis*, and rate above 80%. Microscope observation showed that

基金项目："国家重点研发计划"（2016YFD0201100）；甘肃省国际科技合作专项（1604WKCA010）；甘肃省农牧厅生物技术专项（GNSW-2012-27）

通讯作者：薛应钰，男，汉族，副教授，博士；E-mail：xueyy@gsau. edu. cn

第一作者：杨树，男，满族，硕士；E-mail：yangsaint@126. com。

mycelium was abnormal by *Glycyrrhiza uralensis*, *Albizia julibrissin and Bupleurum chinensis*.

Keywords: Plant extracts; *Valsa mali*; antifungal activity

甘肃省是我国北方主要的水果生产省,同时也是我国苹果生产的第二大省[1-4]。然而,伴随着苹果产业的发展,苹果树腐烂病的危害也日益突出,成为我国苹果产区威胁最大的一种毁灭性病害,严重阻碍了苹果产业的健康高效发展[5,6]。化学药剂防治是目前治理苹果树腐烂病最常用的方法,但化学药剂以及抗生素在为我们的生活带来益处的同时,也存在着污染环境、危害有益生物、易产生抗药性、易在环境和人体内残留,对人类和动物的毒性作用等诸多弊端,已不能很好的适应社会发展。因此新型高效、低毒、低残留、与环境友好的植物性农药的研制与开发已经成为各国研发的热点[7]。

植物提取物是存在于植物体中被人们提取出来的具有生物活性的物质,其有效成分如生物碱、多糖、有机酸、单宁和大蒜素等均具有不同程度的抗菌作用,不但在防治功效上能与化学药剂相媲美,同时还具有能自然分解,对环境和人畜低毒、低残留,对害虫天敌伤害小等优点。因此从天然植物中提取活性物质用于抑制病原菌生长具有广阔的发展前景[8]。

对此已有前人进行研究:孟昭礼等[9,10]发现银杏的乙醇提取液对苹果腐烂病菌有明显的抑制作用;闫红豆等[11]发现侧柏、五倍子、藿香、百部和旋覆花 5 种植物提取物对苹果树腐烂病病菌菌丝的生长也有明显的抑制作用。本研究以苹果腐烂病菌为供试菌株,对 10 个科 15 种植物的提取物进行生物活性测定,发现甘草、柴胡、合欢的乙醇提取物对苹果腐烂病菌均表现出了较强的抑菌活性,值得对其有效杀菌成分进行分离、纯化以及结构鉴定,以期为植物源农药的研制开发提供坚实的理论依据。

1 材料与方法

1.1 材料

1.1.1 供试植物

本试验所采用植物种类见表 1,隶属 10 个科,花椒采于甘肃农业大学,其余药物均购于甘肃省兰州市安宁区药材市场。

表 1 供试植物种类

Table 1 Test plant species

科名 Families	供试植物 Test plant	供试部位 Test site
菊科	千里光(*Senecio scandens*)	全草
	佩兰(*Eupatorium fortunei*)	全草
毛茛科	乌头(*Aconitum carmichaeli*)	根
	白芍(*Paeonia lactiflora*)	根
木兰科	地风(*Cortex illicii*)	茎皮
唇形科	香薷(*Elsholtzia ciliata*)	全草

续表 1

科名 Families	供试植物 Test plant	供试部位 Test site
	薄荷(*Mentha haplocalyx*)	茎皮
	夏枯草(*Prunella vulgaris*)	果实
含羞草科	合欢(*Albizia julibrissin*)	枝皮
豆科	决明子(*Catsia tora*)	果实
	甘草(*Glycyrrhiza uralensis*)	根茎
伞形科	柴胡(*Bupleurum chinensis*)	根茎
百合科	百合(*Lilium brownii* var. *viridulum*)	鳞叶
木通科	通草(*Tetrapanax papyriferus*)	藤木
芸香科	花椒(*Zanthoxylum bungeanum*)	叶

1.1.2 供试植物病原菌

苹果树腐烂病菌(*Valsa mali*)由甘肃农业大学植物保护学院植物病原学实验室提供。

1.2 方法

1.2.1 植物提取液的制备

参考文献[12,13]的方法，将采集的植物阴干，粉碎后过 40 目筛，取 20 g 样品于三角瓶中，加入 75%无水乙醇 100 mL，室温下浸泡 48 h，经 4 层纱布过滤后于 4000 r/min 条件下离心 15 min，取上清液，水浴锅蒸发至干，残余物溶于 20 mL 50%乙醇中，配制浓度为 1000 mg/mL 供试母液，对其进行编号，于 -4℃下保存备用。

1.2.2 植物提取液对病原菌菌丝抑制作用的测定

采用生长速率法[14,15]。将各提取液加入 PDA 培养基中，使其浓度为 10 mg/mL。在平板接入供试菌饼，菌丝一面朝下，每个处理 4 次重复，以加相应量的 50% 乙醇作为对照，置于恒温培养箱(25℃、16 h 光照)中培养。每隔 1 d 用十字交叉法测量菌落直径，至对照长满培养皿为止，并计算抑菌率。计算公式：

$$\text{抑菌率} = \frac{\text{对照直径} - \text{处理直径}}{\text{处理直径} - 5} \times 100\%$$

1.3 数据统计和分析

数据采用 Excel 2007 软件进行统计分析和作图，并采用 SPSS 19.0 软件和 Duncan 新复极差法进行多重比较($P<0.05$)。

2 结果与分析

2.1 15 种植物提取物对苹果树腐烂病菌的抑制作用测定

试验结果表明，15 种植物提取物对苹果树腐烂病菌均有一定的抑制作用。其中豆科植物甘草、含羞草科植物合欢、伞形科植物柴胡的乙醇提取物对苹果树腐烂病菌具有较好抑制率效果，可以有效抑制病原菌菌丝扩展与延伸(图 1 和表 2)。甘草提取物对苹果树腐烂病菌的抑制率最高，72 h 时抑菌率高达 92.95%；合欢和柴胡提取物抑制率达 80%以上。而千里光、夏枯草、薄荷、乌头、白芍、佩兰、百合、通草、花椒对苹果树腐烂病菌的抑制作用相对较低。

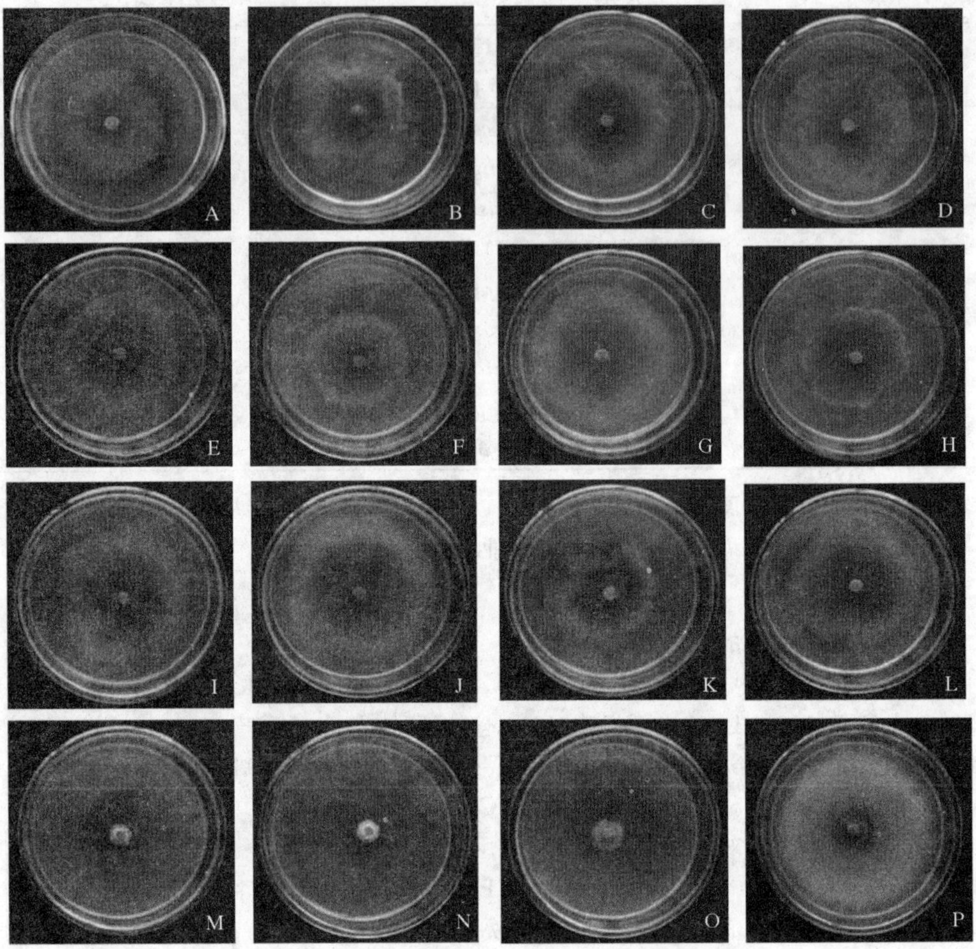

图 1　15 种植物提取物对苹果树腐烂病菌的抑制作用

Fig. 1　Inhibition effect of extracts of 15 kinds of plant against *Valsa mali*

注：A：香薷；B：花椒；C：地风；D：乌头；E：白芍；F：决明子；G：千里光；H：通草；I：佩兰；J：百合；K：夏枯草；L：薄荷；M：合欢；N：甘草；O：柴胡；P：对照。

Note：A：*Elsholtzia ciliata*（Thunb.）Hyland.；B：*Zanthoxylum bungeanum* Maxim.；C：*Schizophragma integrifolium*（Franch.）Oliv.；D：*Aconitum carmichaeli* Debx.；E：*Cynanchum otophyllum*；F：*Cassia tora* Linn；G：*Senecio scandens* Buch.-Ham. ex D. Don；H：*Tetrapanax papyriferus*；I：*Eupatorium fortunei* Turcz.；J：*Lilium brownii* var. *viridulum* Baker；K：*Prunella vulgaris* L；L：*Mentha haplocalyx* Briq.；M：*Albizia julibrissin* Durazz.；N：*Glycyrrhiza uralensis* Fisch；O：*Bupleurum chinensis*；P：Control.

表 2　15 种植物提取物处理后病原菌菌落直径

Table 2　The colony diameter was treated by 15 kinds of plant extracts　mm

供试植物 Test plant	平均菌落直径 Average colony diameter			
	24 h	48 h	72 h	96 h
千里光	16.42	35.86	54.10	73.99
佩兰	12.50	38.92	52.26	65.15

续表 2

供试植物 Test plant	平均菌落直径 Average colony diameter			
	24 h	48 h	72 h	96 h
乌头	13.89	35.16	52.48	70.07
白芍	17.44	36.24	52.42	63.95
地风	11.14	32.11	47.06	66.02
香薷	11.28	29.34	44.98	55.06
薄荷	10.94	31.01	44.26	68.21
夏枯草	17.22	37.83	54.19	58.99
合欢	6.06	9.39	11.17	12.82
决明子	12.19	27.78	48.26	46.11
甘草	7.25	8.52	9.16	11.77
柴胡	7.39	11.09	11.54	14.03
百合	15.53	39.04	61.17	69.07
通草	16.59	32.38	52.08	68.92
花椒	13.22	36.30	58.01	73.76
CK	18.69	42.90	64.034	80.00

2.2 3种植物乙醇提取物对苹果树腐烂病菌菌丝的抑制作用

经显微观察发现，用甘草提取液处理后的苹果树腐烂病菌菌丝呈现为分支增多，菌丝畸形肿胀（图 2-A），柴胡提取液处理后的苹果树腐烂病菌菌丝顶端呈球状膨大，畸形（图 2-B）；合欢提取液处理后的苹果树腐烂病菌菌丝出现原生质浓缩，菌丝断裂现象（图 2-C）；对照菌丝粗细均匀，表面光滑（图 2-D）。

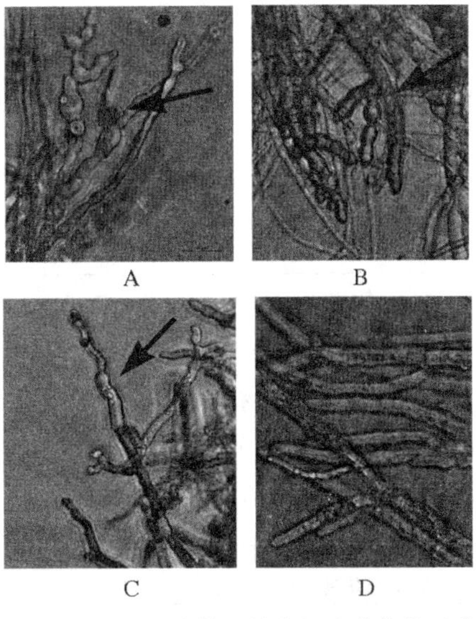

图 2　3种植物提取物对苹果树腐烂病菌菌丝形态的影响

Fig. 2　Effects of three plant extracts on mycelial morphology of *Valsa mali*

A：甘草；B：柴胡；C：合欢；D：对照。

Note：A：*Glycyrrhiza uralensis* Fisch；B：*Bupleurum chinensis*；C：*Albizia julibrissin* Durazz.；D：Control.

3 结论与讨论

本研究通过10个科的15种植物提取物对苹果腐烂病菌的抑制活性进行实验，结果表明15种植物提取物对苹果腐烂病菌都表现出了不同程度的抑制作用。甘草、柴胡、合欢的乙醇提取物均表现出了较强的抑菌活性，对苹果腐烂病菌的抑制率都在80%以上，其中甘草的乙醇提取物的生长抑制率最大，在90%以上。显微形态观察表明，经甘草、柴胡、合欢的乙醇提取物处理后，受抑制菌丝都表现为畸形肿胀，用做开发新型植物源药剂可行性较高(表3)。

表3 15种植物乙醇提取物对苹果树腐烂病菌的抑菌率

Table 3 Antibacterial rate of 15 Kinds of plant extracts on *Valsa mali* %

供试植物 Test plant	抑菌率 Inhibitory rate			
	24 h	48 h	72 h	96 h
合欢	90.27±0.77aA	88.41±0.61bA	89.55±0.63bB	89.58±0.41bA
甘草	89.26±1.84bB	90.71±0.62aA	92.95±0.39aA	90.97±0.44aA
柴胡	82.49±1.44cC	83.93±0.62cB	88.93±0.31bB	87.96±0.77cB
薄荷	56.64±1.96dD	31.39±0.61fE	33.51±1.08cC	15.72±0.32jH
地风	55.18±1.02dD	28.47±1.46gF	28.75±0.99eD	18.64±0.66iG
香薷	54.16±1.76dE	35.78±0.92eD	32.27±1.07dC	33.25±0.87eD
决明子	47.53±1.06eF	39.88±0.92dC	26.72±0.57fE	45.19±0.59dC
佩兰	45.21±1.14eF	10.51±1.32kJ	19.95±0.55gF	19.80±0.57hG
花椒	39.97±1.03fG	17.42±1.07iH	10.21±0.74iH	8.32±0.42lJ
乌头	35.03±2.38gH	20.43±0.96hG	19.58±1.35gF	13.23±0.65kI
百合	23.09±1.63hI	10.18±0.73kJ	4.85±0.87jI	15.31±2.08jH
千里光	16.55±3.33iJ	18.59±3.98iG	16.82±0.89hG	8.01±0.42iJ
通草	15.35±1.41iJ	27.76±1.10gF	20.25±0.38gF	14.78±0.63jH
白芍	9.14±0.62jK	17.58±0.76iH	19.67±0.75gF	21.40±0.53gF
夏枯草	10.75±1.82kL	13.39±1.69jI	16.67±0.35hG	28.01±0.57fE
CK	—	—	—	—

注：同列数据后不同字母表示在0.05水平上差异显著。CK为50%乙醇。

Note: Data in the same column followed by different letters are significantly different at 0.05 level.

鉴于乙醇成本低且相对安全无公害，本研究采用了乙醇作为提取溶剂，会对极性物质的提取能力增强，但不一定能将极性和非极性较强的生物活性物质全部提取出来。溶剂不同会对活性物质的提取效果具有一定影响。对于同种病原菌，植物的乙酸乙酯提取物的抑菌作用高于乙醇提取物的抑菌作用，乙醇提取物的抑菌作用高于石油醚提取植物的抑菌作用[16]。因此，其他有机溶剂提取活性物质的抑菌效果还有待进一步研究。

用不同温度处理植物提取物，有些表现为不受温度影响，而有些植物提取物在较高温度处理后，抑菌效果明显提高；在不同pH条件下，植物提取物对病原菌的抑制效果也有所差异[17]。此外，要确定植物提取物的田间防治效果还必须进行田间试验，以上都还需进一步的研究。

参考文献

[1] 王海波，李林光，陈学森，等. 早熟苹果品种果实的风味物质和风味品质. 中国农业科学，2010，(11)：2300-2306.

[2] 聂继云，李志霞，李海飞，等. 苹果理化品质评价指标研究. 中国农业科学，2012，(14)：2895-2903.

[3] 李国梁. 甘肃省苹果产业现状与发展对策初探. 甘肃农业，2015，(10)：25-26.

[4] 李国梁. 甘肃省苹果产业现状与发展对策. 甘肃农业，2013，(21)：30-32.

[5] 袁正荣，呼丽萍，王秀芳，等. 陇东南地区苹果腐烂病现状调查及防控对策. 甘肃林业科技，2013，(4)：20-24.

[6] 任秀奇，楚景月，刘丽娜，等. 苹果腐烂病综合防控措施. 辽宁林业科技，2010，(4)：50-51，56.

[7] 王静，毛自文，范黎明，等. 30 种植物提取物抑菌活性研究. 安徽农业科学，2012，(10)：5918-5919，6185.

[8] 武月红. 4 种柏科植物提取物抑菌效果的比较研究. 内蒙古农业科技，2015，(2)：25-28.

[9] 孟昭礼，董瑞端，高庆霄. 白果提取液及抑菌作用研究. 农药，1987，(6)：10-15.

[10] 孟昭礼，吴献忠，高庆霄，等. 银杏提取液对四种植物病原菌的抑菌作用. 植物病理学报，1995，(4)：357-360.

[11] 闫红豆，王亚南，唐兴敏，等. 5 种植物提取物抑制苹果树 3 种病原菌活性的研究. 中国果树，2014，(2)：22-26.

[12] Wang S T，Wang X Y，Liu J L，*et al*. Screening of Chinese herbs for the fungitoxicity against *Phytophthora infestan*. Journal of Agricultural University of Hebei，200，24 (2)：101-107.

[13] 闵华，李青松. 中草药提取物对 5 种植物病原菌的抑菌活性研究.安徽农业科学，2009，37(21)：10037-10039.

[14] 吴文君. 植物化学保护试验技术导论. 西安:陕西科学技术出版社，1988.

[15] 何昆，罗宽. 中草药萃取液对植物病原真菌、细菌的抑制作用. 湖南农业科学，2003，(1)：43-45.

[16] 郝妮娜，杨美林，秦小萍，等. 肿柄菊提取物对 7 种植物病原菌的抑菌作用. 中国农学通报，2011，27(21)：272-275.

[17] 曾令达，张荣，蔡韫，等. 17 种植物提取物对荔枝霜疫霉菌的抑制作用. 广东农业科学，2012，(12)：89-92.

代孕体中以同源基因替换为基础的柑橘黄龙病菌致病相关基因的功能研究方法建立

程保平，彭埃天*

（广东省农科院植物保护研究所/广东省植物保护新技术重点实验室，广州 510640）

摘要：柑橘黄龙病已在全球 50 多个国家和地区发生流行，危害巨大，但目前尚缺乏抗病品种和高效药剂，急需研究其致病机制，以寻找新的防控策略。由于其病原菌无法人工培养，本研究回避了这一难题，建立了以可培养的叶缘焦枯病菌为代孕体，通过同源基因替换的方法来研究柑橘黄龙病菌致病相关基因的方法。首先通过生物信息学在柑橘黄龙病菌与叶缘焦枯病菌组中鉴定出一些致病相关的双组分信号转导相关基因，其中一个基因 *ompR* 在两种病原菌中有相似的序列和结构特征。进一步在叶缘焦枯病菌中敲除自身的 *ompR*，得到了 *XfΔompR*（*Xf*）细菌突变体，然后用叶缘焦枯病菌的 *ompR* 基因进行互补，得到了互补突变体 *XfΔompR*－*C*（*Xf*）。野生型、敲除突变体与互补突变体三种菌株在培养基中的生长曲线基本一致。敲除突变体 *XfΔOmpR* 的细菌聚集黏附表型显著减弱、生物膜的形成比野生型菌株减少；接种寄主后该突变体致病性减弱。互补突变体 XfΔ*ompR*－*C*（*Xf*）可回复敲除突变体 *XfΔompR*（*Xf*）的细菌聚集粘附/生物膜形成的表型及相关基因的表达，寄主接种互补突变体 *XfΔompR*－*C*（*Xf*）后可回复致病性。进一步把柑橘黄龙病菌的 *ompR* 基因分别通过同源臂双交换的方法在叶缘焦枯病菌中替换其同源基因，得到了同源替换突变体 *XfΔompR*－*R*（*CLas*）。然后把野生型叶缘焦枯病菌、*XfΔompR*－*C*（*Xf*）和同源替换突变体 *XfΔompR*－*R*（*CLas*）在培养液中进行培养，结果表明：三种菌株在培养液中的生长曲线基本一致。同源替换突变体 *XfΔompR*－*R*（*CLas*）可部分回复敲除突变体 *XfΔompR*（*Xf*）的病原菌聚集黏附及细菌生物膜形成的表型，和相关基因的表达。本研究以同源基因替换为基础，初步在叶缘焦枯病菌代孕体中建立了黄龙病菌致病相关基因的功能研究方法。

关键词：柑橘黄龙病；双组分信号转导基因；致病性；同源基因替换

基金项目：广东省重大科技计划专项（2014B020203003）；广东省自然科学基金（2014A030310198）；广东省现代农业产业技术体系建设专项（2016LM1077）；广东省科技计划项目（广东省农作物病虫害绿色防控技术研究开发中心建设）；广东省农业标准化项目（粤财农[2016]310 号）；广东省科技计划重点实验室项目（2014B030301053 ）

通讯作者：彭埃天（1962-　），男，研究员；E-mail：pengait@163. com

第一作者：程保平（1982-　），男，助理研究员；E-mail：1093449912@qq. com。

1 材料与方法

1.1 细菌生长曲线测定

在 50 mL 的聚丙烯管中加入 30 mL PD_2 培养液，然后接种野生型 *Xylella fastidiosa* 和突变体菌株，置于 28℃ 培养箱中，150 r/min 振荡培养 10 d，每天定时取 1 mL 培养液检测 OD_{600} nm 值。试验重复 3 次。

1.2 细菌聚集测定

在 50 mL 的聚丙烯管中加入 15 mL PD_2 培养液，然后接种野生型和突变体菌株，在 28℃ 培养箱中垂直放置 10 d。此时培养液的上层是充分分散的菌体溶液，OD_{540} nm 读数值可更好地反映该细菌的聚集状态，测量上层菌液 OD_{540} nm 的吸光值 ODs。然后把吸光度测量杯中的菌液吸回培养管，在振荡器上振荡 1 min，使菌体充分分散，然后马上测量培养液的 OD_{540} nm 吸光值 ODt。细菌聚集的相对百分比使用如下公式计算：(ODt-ODs)/ODt×100，试验重复 3 次。

1.3 细菌生物膜形成

在 96 孔聚苯乙烯微孔培养板中每孔加入 100 μL PD_2 培养液，每孔接种 10 μL 菌液(野生型或突变体)，28℃ 静置孵育 10 d，然后将培养液吸出，每孔加入 100 μL 无菌 PBS 缓冲液清洗板孔 3 次。接着每孔加入 100 μL 甲醇固定 15 min，然后吸出培养孔中的甲醇，自然风干，随后每孔加入 100 μL 的 1%结晶紫溶液，室温下染色 15 min，然后吸出培养孔中的结晶紫染色液，用超纯水把多余的染料冲洗干净，把培养板倒置在滤纸上除去残余的超纯水，每孔加入 100 μL 乙醇溶液，在 37℃ 培养箱中作用 30 min 以溶解结晶紫，最后在 590 nm 条件下，测定培养孔中溶液的 OD 值。每次试验每种菌株做 8 个孔的重复，OD 试验数值取平均值。试验重复3 次。

1.4 基因敲除/互补/同源臂双交换

参照实验室前期论文，首先扩增目标基因的上下游片段与抗性基因片段，然后用 Overlap-PCR 的方法连接三个片段，形成一个抗性基因两侧具有与目标基因上下游同源序列的靶标基因替换盒，接着克隆到自杀质粒中，用电击转化的方法把质粒转入感受态细菌，电转后将细菌涂布于抗性平板，此时上下游同源臂可与宿主染色体中的目标片段发生两次同源交换，使替换盒整合到宿主靶标处，使寄主可在抗性平板上生长，单菌落长出后用基因引物及抗生素引物进行鉴定。

1.5 致病性测定

野生型与突变体病原菌在 PD_2 琼脂平板上 28℃ 生长 5 d，用超纯水悬浮，并调整 OD_{600} 至 0.1，然后取 20 μL 菌液，参照前人对的方法用接种针接种寄主(1)，每种菌株接种 10 棵寄主，并设置一个清水接种对照，植物放置于温室 24～32℃，18 h/6 h 光暗交替培养。4 个月后，统计分析病株的病情指数，并取病样研磨，用荧光定量法检测病样中的病原菌浓度。

2 结果与分析

利用生物信息学在柑橘黄龙病菌基因组中筛选出了一个致病相关的双组分信号转导相关基因：*ompR*，如图 1 所示。

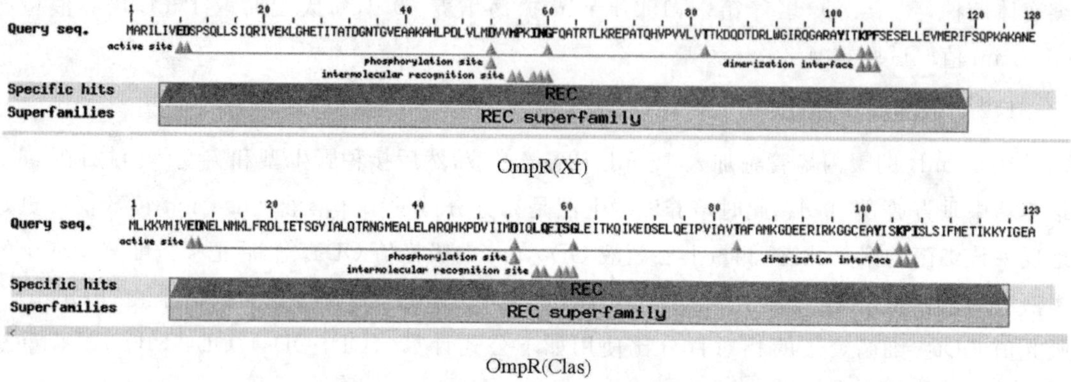

图 1 蛋白 ompR 在柑橘黄龙病菌与叶缘焦枯病菌中均有相似的序列和结构特征

柑橘黄龙病菌的 *ompR* 基因在叶缘焦枯病菌的基因组中存在同源基因，叶缘焦枯病菌的 *ompR* 基因敲除实验表明：其在细菌聚集黏附、生物膜形成及致病过程中发挥重要作用（图 2 至图 7）。

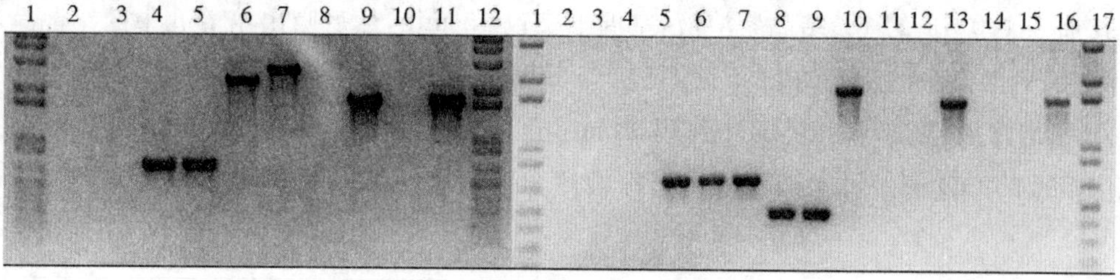

图 2 突变体插入片段验证

左图泳道 2,3：*Xf*(WT)和 Xf*ΔompR*(*Xf*)的空白检测对照；泳道 4,5：用叶缘焦枯病菌的特异基因分别检测：*Xf*(WT)和 *XfΔompR*(*Xf*)；泳道 6,7：用同源臂上下游引物分别检测：*Xf*(WT)和 *XfΔompR*(*Xf*)；泳道 8,9：用同源臂上游引物 + 抗生素片段下游引物分别检测：*Xf*(WT)和 *XfΔompR*(*Xf*)；泳道 10,11：用同源臂下游引物 + 抗生素片段上游引物分别检测：*Xf*(WT)和 *XfΔompR*(*Xf*)。右图泳道 2,3,4：*Xf*(WT)，*XfΔompR*(*Xf*)和 *XfΔompR-C*(*Xf*)的空白检测对照；泳道 5,6,7：用叶缘焦枯病菌的特异基因分别检测 *Xf*(WT)，*XfΔompR*(*Xf*)和 *XfΔompR-C*(*Xf*)；泳道 8,9,10：用插入区段上下游引物分别检测 *Xf*(WT)，*XfΔompR*(*Xf*)和 *XfΔompR-C*(*Xf*)；泳道 11,12,13：用插入区段上游引物 + *ompR* 基因下游引物分别检测：*Xf*(WT)，*XfΔompR*(*Xf*)和 *XfΔompR-C*(*Xf*)；泳道 14,15,16：用用插入区段下游引物 + *ompR* 基因上游引物分别检测：*Xf*(WT)，*XfΔompR*(*Xf*)和 *XfΔompR-C*(*Xf*)。

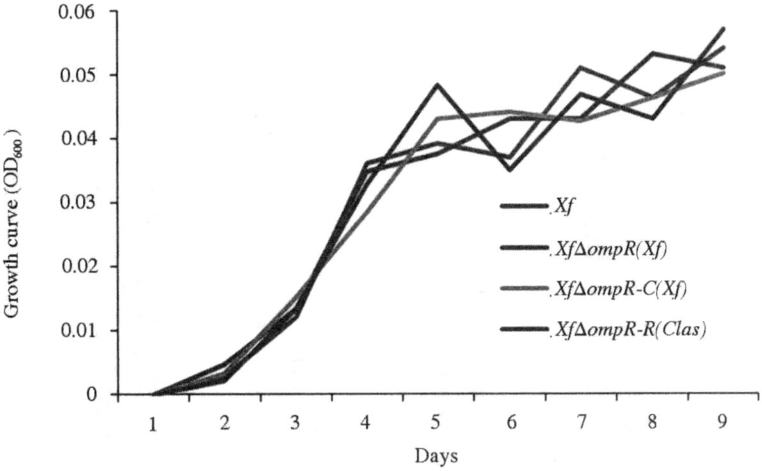

图 3　3 个突变体菌株的生长曲线与野生型一致

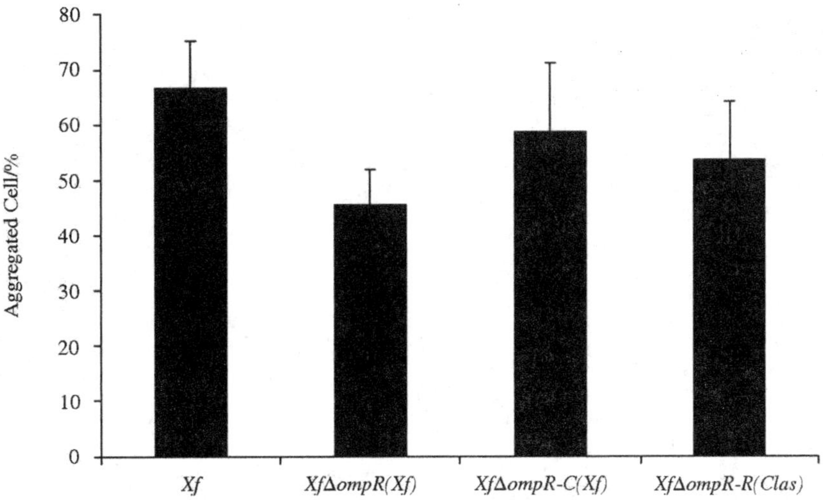

图 4　敲除叶缘焦枯病菌的 *ompR* 基因后,突变体 *XfΔompR* 的细菌聚集表型显著减弱,而回复突变体 *XfΔompR-C*(*Xf*)及同源基因替换突变体 *XfΔompR-R*(*CLas*)均可回复敲除突变体 *XfΔompR* 的病原菌聚集的表型

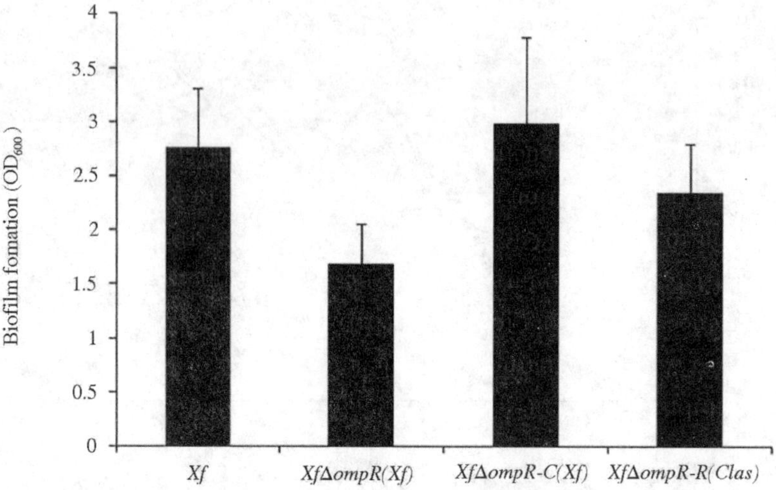

图 5 图 4 敲除叶缘焦枯病菌的 *ompR* 基因后，突变体 ***XfΔompR*** 的生物膜形成表型显著减弱，而回复突变体 ***XfΔompR-C(Xf)*** 及同源基因替换突变体 ***XfΔompR-R(CLas)*** 均可回复敲除突变体 ***XfΔompR*** 的细菌生物膜形成的表型

图 6 ***XfΔompR-C(Xf)*** 及同源基因替换突变体 ***XfΔompR-R(CLas)*** 均可互补其同源基因 *ompR* 的功能，回复致病性

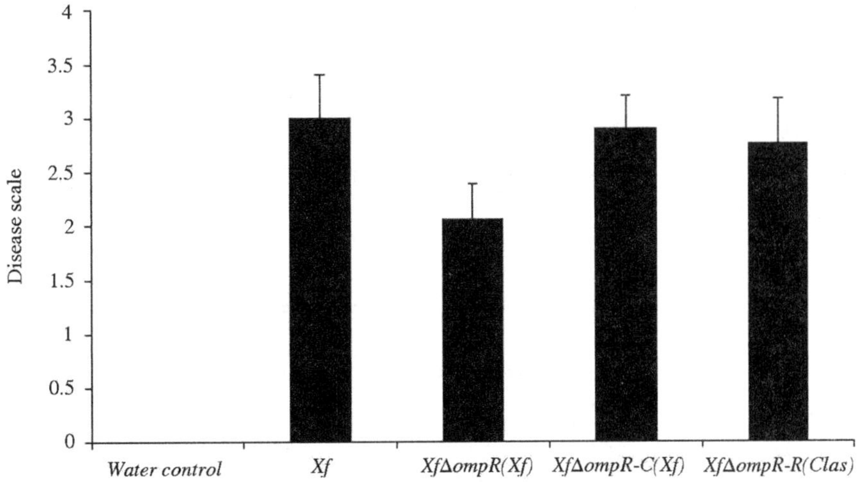

图 7 *XfΔompR-C*(*Xf*)及同源基因替换突变体 *XfΔompR-R*(*CLas*)均可互补其同源基因 *ompR* 的功能，回复致病性

参考文献

Shi X Y, Dumenyo C K, Hernandez-Martinez R, *et al*. Haracterization of regulatory pathways in *Xylella fastidiosa*: genes and phenotypes controlled by *gacA*. Appl Environ Microbiol, 2009, 75: 2275-2283.

First report of brown leaf spot disease in kiwifruit in China caused by *Neofusicoccum parvum*

Li Li#, Hui Pan#, Meiyan Chen, Shengju Zhang, Caihong Zhong*

(*Key Laboratory of Plant Germplasm Enhancement and Specialty Agriculture, Wuhan Botanical Garden, Chinese Academy of Sciences, Wuhan 430074, China*)

Abstract: Kiwifruit (*Actinidia* sp.) is economically important worldwide. Its production has increased hugely over the last 30 years in China, New Zealand, Italy and Chile[1]. In August 2016, brown leaf spot disease was found in orchards in Liupanshui, Guizhou Province and Taishun, Zhejiang Province. Symptoms appeared as foliar discolorations, such as grey to brown sectorial necrosis, with infected leaves dying while attached to the stem (Fig. 1-A, B, C). To isolate the causal agent, diseased leaf tissues were surface-disinfected with 70% ethanol, then 1% sodium hypochlorite for 1 min, rinsed twice with sterile distilled water and plated onto Petri dishes with potato dextrose agar (PDA). Hyphal tip from the growing edge of colonies cultured for 3 days at 25°C was transferred to PDA to obtain pure isolates. After 5 days at 25°C, ten morphologically similar fungal isolates (NP1-10) were collected from each location. Fungal colonies were white with aerial mycelium, then turned dark green within two week of incubating plates at 25°C (Fig. 1-D, E). Conidia were fusoid, hyaline, unicellular with sub-obtuse apex and truncate base, fusiform to ellipsoidal with rounded apices, truncated bases, thin walls, and a smooth surface. Fifty conidia were measured and ranged from 15.43～17.58 μm in length and 4.37～6.99 μm in width, with a L/W ratio of 3.2. These traits match those reported for *Neofusicoccum parvum*[2,3]. To confirm the identities of ten isolates, the rDNA internal transcribed spacer (ITS) region and β-tubulin (BT) gene were amplified and sequenced by PCR using primers (ITS4 and ITS5 and BT2a and BT2b). The sequences (Accession no. MF314146-MF314155

基金项目：国家自然科学基金青年科学基金项目(31701974)；湖北省自然科学基金面上项目(2017CFB443)；湖北省技术创新专项重大项目(2016ABA109)

通讯作者：钟彩虹，研究员，猕猴桃等果树的种质资源保护、鉴定、遗传育种、发育生理和产业技术研究；电话：13407147939，E-mail：zhongch1969@163.com

共同第一作者：李黎，博士，助理研究员，猕猴桃抗性机理研究；E-mail：lili@wbgcas.cn

潘慧，硕士，助理工程师，猕猴桃病害检测。

for the ITS region; MF314156 and MF314165 for β-tubulin) were identical to ex-type *N. parvum* strain MAR2134 (Accession no. KU997474 and KU997593) reported in GenBank using BLASTn, confirmed by molecular phylogenetic trees constructed using MEGA7 (Fig. 2). The species was thus identified as *N. parvum*. A pathogenicity test was performed: a conidial suspension (10 μL, 10^6 conidia/mL) and colonized PDA pieces (5 mm in diameter) from 7 day-old cultures of the fungus in Petri dishes were inoculated on 12 non-wounded/wounded young leaves of cultivar 'Hongyang'. They were then incubated in a controlled greenhouse at 23℃ with 80%～85% humidity for a 12 h photoperiod. Sterile PDA plugs and distilled water given to an equal number of plants with the same methods served as controls. The test was repeated twice. Each wounded inoculated plant developed necrotic lesions like those in the field 5 days after inoculation, while control and unwounded leaves remained symptomless. *N. parvum* was consistently re-isolated from the lesions of artificially inoculated experimental leaves, but no symptoms developed on any control treatments, confirming Koch's postulates. The results confirm *N. parvum* as the causal agent of kiwifruit brown spot disease in China. *N. parvum* has been reported to cause dieback and canker in various fruit trees, including blueberry[4] and mango trees[5]. This is the first known report of *N. parvum* causing brown spot disease in kiwifruit in China.

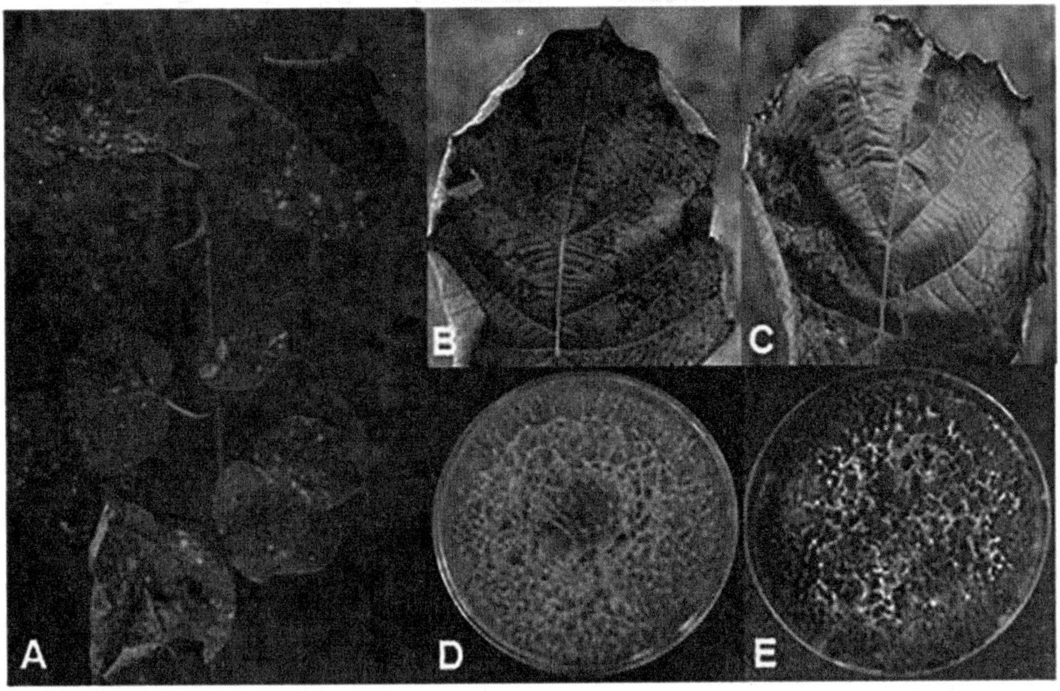

Fig. 1 Natural symptoms of kiwifruit brown leaf spot disease and morphological characteristics of *Neofusicoccum parvum* (A-C) Typical symptoms of naturally-infected leaves of the cultivar 'Hongyang'; (D) Mycelial colony of a 7 day-old culture of *N. parvum* on PDA at 25°C; (E) Mycelial colony of a 20 day-old culture of *N. parvum* on PDA at 25°C.

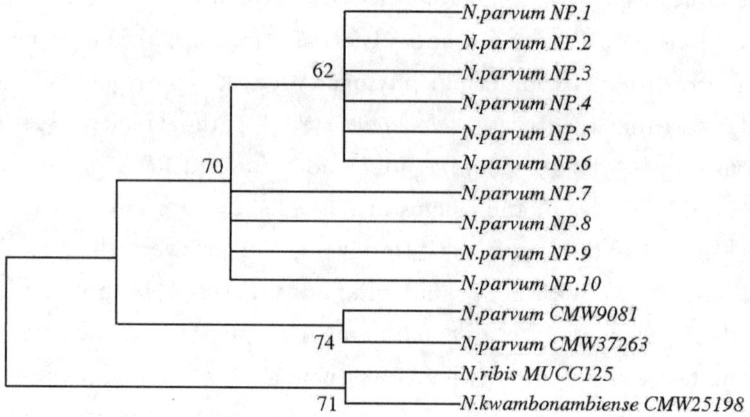

Fig. 2 Molecular phylogenetic tree based on combined ITS and BT sequence alignment generated from a Maximum Likelihood analysis. Values above the branches represent bootstrap support values (>50%).

Fang X L, Finnegan P M, Barbetti M J. Wide variation in virulence and genetic diversity of binucleate Rhizoctonia isolates associated with root rot of strawberry in Western Australia. PLoS ONE, 2013, 8(2): 1-14.

References

[1] Belrose Inc. World kiwifruit review. Washington, USA. 2016, 14.

[2] Crous P W, Slippers B, Wingfield M J, et al. Phylogenetic lineages in the *Botryosphaeriaceae*. Studies in Mycology, 2006, 55: 235-254.

[3] Palavouzis S C, Tzamos S, Paplomates E, *et al*. First report of *Neofusicoccum parvum* causing shoot blight of pomegranate in Northern Greece. New Disease Reports, 2015, 32:10.

[4] Castillo S, Borrero C, Castano R, *et al*. First report of canker disease caused by *Neofusicoccum parvum* and *N. australe* on blueberry bushes in Spain. Plant Disease, 2013, 97: 1112.

[5] Serratodiaz L M, Perezcuevas M, Riveravargas LI, *et al*. First report of *Neofusicoccum parvum* causing rachis necrosis of mango (*Mangifera indica*) in Puerto Rico. Plant Disease, 2013, 97: 1381.

Peanuts of galacturonic acid polymer enzyme inhibition protein gene cloning and expression analysis of *HyPGIP*

Yuxuan Zhou[1], Yidan Shen[1], Yan Wang[2], Zhikui Wang[3], Ruqin Zhang[1], Yicheng Chi[4], Honghai Yan[1*]

([1] *College of Agronomy and Plant Protection, Qingdao Agricultural University, Qingdao 266109, China*;
[2] *Shandong rushan city agriculture bureau, Weihai 266100, China*;
[3] *Jimo agriculture bureau, Qingdao 266200, China*;
[4] *Peanut research institute in shandong province, Qingdao 266101, China*)

Abstract: Peanut leaf rot is a kind of new diseases, we produce the cell wall degradation enzymes found in previous research is an important pathogenic factor, pathogen especially of galacturonic acid polymer enzyme activity is closely related to the pathogenic capability, this study aims to through the specialization of galacturonic acid polymer enzyme inhibition protein PGIP gene expression analysis and research, to parse the PGIP gene function, and then discusses the host disease resistance mechanisms and new ways of peanut leaf rot prevention. By measuring the peanut plants within plant after inoculated leaf rot pathogen of galacturonic acid enzyme activity change and the polymer of galacturonic acid polymer enzyme inhibition protein generation changes, to study how galacturonic acid enzyme in peanut leaf rot diseases pathogenic role and the role of inhibition of protein in the host, disease resistance; According to the GENEBANK database of five peanut PGIP gene mRNA sequence of CDs area (no introns), the design of primers by total RNA extraction, reverse transcription to cDNA, PCR amplification, purpose stripe recycling connection T carrier, purified and transformation of e.c. with our fabrication: oli DH 5 alpha, get peanuts HyPGIP gene cloning, sequencing and ORF Finder analysis HyPGIP gene sequences, predict HyPGIP basic properties and biological function of protein. Through in peanut plant inoculated leaf rot pathogen induced HyPGIP gene expression, to analyze HyPGIP protein and disease-resistant relationship and action mechanism of peanut. Result: Disease-

基金项目：山东省自然科学基金项目(ZR2011CL005)；山东省“泰山学者”建设工程专项经费资助(BS2009NY040)
通讯作者：鄢洪海(1964-)，男，吉林长春人，教授，博士，主要从事植物病理学与玉米病害研究；E-mail: hhyan@qau.edu.cn
共同第一作者：周宇轩(1993-)，女，山东潍坊人，在读本科生，主要从事植物病害防治研究；E-mail: 814532720@qq.com
沈怿丹同等贡献。

resistant varieties breeding bacteria inoculated leaf rot 25 blades after disease spot is smaller than baisha 1016 cultivars, polymer galacturonic acid enzyme activity beginning to rise, but soon drops, and white 1016 poly galacturonic acid enzyme activity significantly more weak. And generate the amount of enzyme inhibition protein of galacturonic acid polymer baisha 1016 significantly more than, that of galacturonic acid polymer enzyme activity closely associated with peanut leaf rot pathogen pathogenicity, and enzyme inhibition protein of galacturonic acid polymer is closely related with the host disease resistance. HyPGIP obtained with gene cloning and sequencing of gene sequence length is 1427 bp, the gene sequence contains a 1029 bp complete ORF, encoding 342 amino acids, no introns, termination codon TGA, named HyPGIP gene, the ORF sequences with the reported five peanut PGIP sequence homology of ORF area is high; Structure prediction found eight LRR structure found in the sequence domain, signal peptide, hydrophobic strong; After inoculation host bacteria treatment, amount of PGIP gene expression in the peanut plants increased, and with the extension of inoculation time and rising, and expressed in 36 to 48 h period quantity increases rapidly, and then changes slowly decrease. Rt-pcr test showed that in the normal, healthy plants, PGIP gene expression quantity is little, but increased after inoculated leaf rot pathogen, in 36 within 48 h-a rapidly increasing process, after generation slowly decrease, shows affected by the pathogen infect HyPGIP gene expression, also is not a uniform increase process.

Verticillium dahliae LHS1 is required for virulence and expression of extracellular enzymes involved in cell-wall degradation

Weiye Cui[1], Dandan Zhang[1], Jun Zhang[1], Jieying Chen[1], Xiaofeng Dai[1,2*], Wei Guo[1,2*]

([1] *Institute of Food Science & Technology, Chinese Academy of Agricultural Sciences, Beijing 100193, China*; [2] *Key Laboratory of Agro-products Quality and Safety Control in Storage and Transport Process, Ministry of Agriculture, P. R. China*)

Abstract: *Verticillium dahliae* is a devastating soilborne fungus causing vascular wilt disease in a wide range of economically important crops. Cell wall-degrading enzymes (CWDE) secreted by *V. dahliae* have been considered to play important roles in pathogenesis. However, little is known about the secretion mechanisms of CWDE used by this pathogen. In this study, we investigated the role of the *V. dahliae LHS1* (*VdLHS1*), the homolog of which in *Magnaporthe oryzae* is an endoplasmic reticulum (ER) chaperone protein that was previously implicated in pathogenicity and expression of extracellular enzymes. We showed that *VdLHS1* colocalized with ER-tracker dye Blue-White DPX in ER. The *ΔVdLHS1* mutant showed defects in radial growth and asexual conidiospores. The disease index of cotton inoculated with *ΔVdLHS*1 mutant was significantly lower than that of wildtype strain throughout the assay up to 30 days post-inoculation. On complete medium (CM) plates supplemented with hydrogen peroxide, sodium chloride, sorbitol and SDS, the growth rates of the *ΔVdLHS1* mutant and the wildtype were not significantly different from those on CM. Furthermore, the activities of several extracellular enzymes in culture filtrates from the *VdLHS1* mutant and the wildtype were compared using different test plate assay and enzyme linked immunosorbent assay. The activities of celluase, ligninase, laccase, pectinase and xylanase were significantly reduced in the *ΔVdLHS1* mutant, and those of amylase and beta-glucosidase did not change markedly in the mutant. However, activities of protease increased remarkably in the *ΔVdLHS1* mutant. Together, our results suggest that the *VdLHS1* is requisite for pathogenicity, conidiation and secretion of extracellular enzymes.

Corresponding authors: Email: daixiaofeng@caas.cn; guowei01@caas.cn, Tel: +86-10-62813566; Fax: +86-10-62813566.

Molecular cloning and functional analysis of two novel polygalacturonase genes in *Rhizoctonia solani*

Xijun Chen[1*], Lili Li[1], Zhen He[1*], Jiahao Zhang[1], Benli Huang[1], Zongxiang Chen[2], Shimin Zuo[2], Jingyou Xu[1]

([1] *Horticulture and Plant Protection College, Yangzhou University, Yangzhou 225009, China;*
[2] *Key Laboratory of Crop Genetics and Physiology/Co-Innovation Center for Modern Production Technology of Grain Crops, Key Laboratory of Plant Functional Genomics of the Ministry of Education, Yangzhou University, Yangzhou 225009, China*)

Abstract: Two novel polygalacturonase (PG) genes, *RsPG*3 and *RsPG*4, were cloned from *Rhizoctonia solani* isolate YN-7, which causes rice sheath blight. The predicted protein product of RsPG3 contained 239 amino acid (aa) residues with a molecular mass of 25.0 ku, whereas *RsPG4* encoded a deduced protein of 345 aa residues with a mass of 37.5 ku. Sequence alignment, phylogenetic analysis and gene ontology indicated that RsPG3 and RsPG4 had endo-PG and exo-PG activity, respectively. Recombinant RsPG3 and RsPG4 both exhibited PG activity that led to the destruction of rice sheaths and release of reducing sugars. Pathogenicity assays showed that the two PGs induced tissue necrosis in rice sheaths, indicating that both RsPG3 and RsPG4 are important virulence factors in the *R. solani*/rice interaction. Further studies are underway to more clearly define their role in rice cell wall degradation and their interaction with PG inhibitor proteins.

Keywords: Polygalacturonase; *Rhizoctonia solani*; functional analysis; rice sheath blight

Integrated transcriptome and metabolome analysis of *Oryza sative* L. upon infection with *Rhizoctonia solani*, the causal agent of rice sheath blight

Wenlei Cao[1], Bin Xu[1], Limin Yuan[2], Xijun Chen[1], Xuebiao Pan[1], Shimin Zuo[1]*

([1] Jiangsu Key Laboratory of Crop Genetics and Physiology/Co-Innovation Center for Modern Production Technology of Grain Crops, Yangzhou University, Yangzhou 225009, China; [2] Testing center of Yangzhou University)

Abstract: Sheath blight (SB), caused by necrotrophic fungus *Rhizoctonia solani* Kühn, gives rise to severely grain losses in rice. In this study, an integrated approach of transcriptomics and metabolomics was employed to try to uncover potential mechanism and to mine the candidate genes/pathways regulating rice SB resistance. YSBR1 is a novel germplasm with stable and high resistance to SB, which has been identified by using greenhouse and field trials. Histological analysis by SEM showed that the growth of *R. solani* mycelium was apparently suppressed in YSBR1 but not in Lemont (susceptible) after 60 HAI. *R. solani* infection was able to trigger substantial metabolic reprogramming in rice, particularly in primary metabolism and energy metabolism. Phenylalanine and tyrosine, which have been confirmed to involve in plant-pathogen interaction via the biosynthesis of secondary metabolites, were specially increased in YSBR1 infected. Total of 332/62 (Lemont/YSBR1) and 2606/748 (Lemont /YSBR1) DEGs at 10 and 20 HAI were identified, respectively. A total of 247 unique DEGs were identified specifically in YSBR1, which were annotated to involve in several pathways, like amino acid metabolism, secondary metabolic, energy metabolism, photosynthesis, and sugar metabolism pathways. Through combining genetic analysis and omics evidence, 6 candidate genes localized within the known SBR QTL region (Lemont/YSBR1), encoding cellulose synthase-like family A, transcription factors TFⅡS, elongin A, EF hand family protein, calmodulin-related calcium sensor protein, ERF transcription factor, and thaumatin, have been identified.

Keywords: *Rhizoctonia solani*; rice; metabolome; transcriptome; QTLs; candidate genes/ pathways

Corresponding author: Shimin Zuo; E-mail: smzuo@yzu. deu. cn.

FRG3, a target of slmiR482e-3p, actives guardians against the fungal pathogen *Fusarium oxysporum* in tomato

Huimin Ji[1#], Min Zhao[1#], Ying Gao[1], Xinxin Cao[1], Huiying Mao[1], Yi Zhou[1], Wenyu Fan[1], Shouqiang Ouyang[1,2,3*], Peng Liu[4*]

([1] *College of Horticulture and Plant Protection, Yangzhou University, Yangzhou 225009, China;*
[2] *Joint International Research Laboratory of Agriculture and Agri-Product Safety, Yangzhou University, Yangzhou 225009, China;*
[3] *Key Laboratory of Plant Functional Genomics of the Ministry of Education, Yangzhou University, Yangzhou 225009, China;*
[4] *Testing Center, Yangzhou University, Yangzhou 225009, China*)

Abstract: Majority plant disease resistance (*R*) genes encode nucleotide binding site-leucine-rich repeat (NBS-LRR) protein specifically determining the plant immune response, which are revealed as targets of several miRNA families during the past decades. The fungus *Fusarium oxysporum* f. sp. *lycopersici* (FOL) causes vascular wilt disease in tomato worldwide. In this study, we explored a possible role for *FGR3* in tomato defense against FOL. *FRG3* was predicted as a *NBS-LRR* like gene, and targeted by slmiR482e-3p which is a member of slmiR482 family. Our Northern blot data demonstrated that all seven members of slmiR482 family were regulated by the infection of FOL in different pattern. The ability of *FRG3* to be regulated by the slmiR482e-3p was confirmed at transcript level by co-expression in *Nicotiana benthamiana*. A virus-induced gene silencing (VIGS) approach revealed that *FRG3* rendered resistance to the resistant tomato Motelle. Together, our study identified a novel *R* gene *FRG3* conferring tomato wilt disease via targeted by corresponding slmiR482e-3p at transcript level *in planta*.

Bioinformatic identification and transcriptional analysis of *Colletotrichum gloeosporioides* candidate CFEM effector proteins

Liqing Zhang, Ke Duan, Chengyong He, Qinghua Gao*

(Forestry and Fruit Tree Research Institute, Shanghai Academy of Agriculture Science, Shanghai 201403, China)

Abstract: Anthracnose caused by *Colletotrichum* spp. is a devastating disease of cultivated strawberry (*Fragaria* × *ananassa* Duch). *Colletotrichum gloeosporioides* is one of the causal agents. Plant pathogens secrete effector proteins into host tissues to promote infection through the manipulation of host processes. CFEM(common in several fungal extracellular membrane proteins)domain is unique to fungi and is associated with pathogenicity. Some of the fungal effectors which have been reported contain the CFEM domain. In this study, we first combined bioinformatics approaches to determine the repertoire of CFEM effector encoded in the *C. gloeosporioides* genome and conducted expression analyses on selected candidates. The results showed that there were 22 *C. gloeosporioides* CFEM proteins were identified and only 8 of them were secreted proteins. These 8 secreted proteins revealed diverse expression patterns, among them, one expressed in in-planta appressorium stage, two expressed in biotrophic stage and one expressed in necrotrophic stage. Finally, we identified 8 effector candidates based on these properties. Our findings will facilitate the functional analysis of fungal pathogenicity determinants and should prove useful in the search for plant quantitative disease resistance components active against the strawberry anthracnose.

Keywords: *Colletotrichum gloeosporioides*; effector; CFEM protein

Corresponding author: Qinghua Gao; E-mail: qhgao20338@sina.com.

A receptor like kinase gene *SBRR1* positively regulates resistance to sheath blight in rice

Zhiming Feng, Yafang Zhang, Yu Wang, Huimin Zhang, Wenlei Cao, Zongxiang Chen, Xuebiao Pan, Shimin Zuo*

(*Key Laboratory of Crop Genetics and Physiology/Co-Innovation Center for Modern Production Technology of Grain Crops, Key Laboratory of Plant Functional Genomics of the Ministry of Education, Yangzhou University, Yangzhou 225009, China*)

Abstract: Sheath blight (SB), caused by the necrotrophic fungal pathogen *Rhizoctonia solani* Kühn, poses a great threat to the rice grain yield as one of the most serious rice diseases. Receptor like kinases play important roles in plant defense responses to disease. However, receptor like kinases associated with rice SB have not been reported. We analyzed the differential gene expression profiles of the rice variety YSBR1 with high resistance to SB before and after inoculation with *R. solani*, and isolated a rice receptor like kinase gene *SBRR1* (*sheath blight-related RLK gene 1*) induced by *R. solani*. *SBRR1* is preferentially expressed in leaf sheaths and leaves, which is consistent with the main infective site of *R. solani*, implying its role in regulating rice resistance to SB. The *SBRR1* loss-of-function mutant *sbbr1* is more susceptible to SB compared with its wild type. Moreover, knockdown of *SBRR1* expression by RNAi reduces the resistance of rice to SB. These results demonstrated that *SBRR1* positively regulates resistance to SB. We further identified an *SBRR1*-interaction protein SIP1 (*SBRR1* interaction protein 1) that encodes an ankyrin repeat protein via yeast two hybrid screening. SIP1 is also induced by R. solani and preferentially expressed in leaf sheaths and leaves. Our results enrich the knowledge of the molecular mechanism of resistance to rice SB, which will provide the theories basis for establishing a new strategy to prevent and control this disease.

Keywords: rice sheath blight; receptor like kinase; *SBRR1*; resistance mechanism

Corresponding author: Shimin Zuo; E-mail: smzuo@yzu.edu.cn.

A sheath blight resistance QTL *qSB-11*^{LE} encodes a receptor like protein involved in rice early immune response to *Rhizoctonia solani*

Xiang Xue, Yu Wang, Zhiming Feng, Huimin Zhang, Wenfeng Shan, Wenlei Cao, Yafang Zhang, Zongxiang Chen, Xuebiao Pan, Shimin Zuo*

(Key Laboratory of Crop Genetics and Physiology/Co-Innovation Center for Modern Production Technology of Grain Crops, Key Laboratory of Plant FunctionalGenomics of the Ministry of Education, YangzhouUniversity, Yangzhou 225009, China)

Abstract: Sheath blight (SB) is one of the three most serious diseases, which leads to severe grain loss annually. Rice resistance to SB disease is a typical of quantitative trait controlled by polygenes or quantitative trait loci (QTLs). Lots of QTLs for SB resistance have been reported while none of them was isolated so far, which significantly hinders the elucidation of resistant mechanism. Previously, we have mapped an SB resistance QTL on chromosome 11 into a 78 kb region, in which the resistant allele (here after called *qSB-11*^{LE}) is from Lemont (LE). Here, through a series of experiments like RNA interference, overexpression and complementation tests, we confirmed that a receptor-like protein lack of intracellular domain in the fine mapping region is *qSB-11*^{LE}. RT-PCR data showed that *qSB-11*^{LE} transcription was significantly induced by SB pathogen *Rhizoctonia solani* but not by rice blight pathogen *XOO*. The *qSB-11*^{LE} is highly expressed in leaf sheath at the booting stage. *Subcellular localization* assay indicated that *qSB-11*^{LE} protein localizes in both *nucleus* and membrane. Through the treatments of known pathogen associated molecular patterns (PAMPs), chitin and flg22, we found that knock down of *qSB-11*^{LE} significantly affected rice innate immune response to chitin. Some known receptor/co-receptor like kinase proteins were found no interaction with *qSB-11*^{LE} protein in *vitro*. Upon the infection of SB fungus, the synthesis of hormone ethylene (ET) was apparently induced in wild plants, while it was only weakly increased in *qSB-11*^{LE} RNAi plants as well as in susceptible near isogenic line (NIL). Spray of ET was able to recover the resistance phenotype of *qSB-11*^{LE} RNAi and NIL plants. Collectively, we conclude that *qSB-11*^{LE} is involved in rice early immune response to SB fungus and regulates ET synthesis upon SB pathogen infection to activate ET-dependent defense response.

Keywords: rice sheath blight; QTL; *qSB-11*^{LE}; receptor like protein; early immune response; ethylene

Corresponding author: Shimin Zuo; E-mail: smzuo@yzu.edu.cn.

A single amino-acid substitution of polygalacturonase inhabiting protein (OsPGIP2) allows rice to acquire higher resistance to sheath blight

Xijun Chen[1], Lili Li[1], Lina Zhang[1], Bin Xu[1], Zhen He[1], Zongxiang Chen[2], Shimin Zuo[2], Jingyou Xu[1]

([1] *Horticulture and Plant Protection College, Yangzhou University, Yangzhou 225009, China;*
[2] *Key Laboratory of Crop Genetics and Physiology of Jiangsu Province/Key Laboratory of Plant Functional Genomics of the Ministry of Education, College of Agriculture, Yangzhou University, Yangzhou 225009, China*)

Abstract: A possible strategy to control plant diseases is improvement of natural plant defense mechanisms against the infection of pathogens. As one kind of leucine-rich repeat proteins, polygalacturonase inhibiting proteins (PGIPs) have been found involved in plant defense against the pathogens by inhibiting the activity of fungal endo-polygalacturonases (endo-PGs). To dig more resistant gene, a single amino acid of the OsPGIP2, which has no effect on the endo-PGs activity of *Rhizoctonia solani*, was substituted and it allowed OsPGIP2 to acquire the inhibiting ability. Overexpression of *ospgip2* significantly increased rice resistance to rice sheath blight and counteracted the tissue degradation caused by RsPGs. Furthermore, overexpression of *ospgip2* did not affect rice agronomic traits or yield components. All these indicated that *ospgip2* was an important potential gene that could be applied to rice sheath blight resistance breeding.

Keywords: single amino acid substitution; polygalacturonase inhibiting protein; sheath blight resistance; *Oryzae sativa*

Understanding the mechanism underlying wheat resistance to Fusarium head blight (FHB) by integrating multi-omics

Yao Zhou, Lei Li, Tao Li*

(Yangzhou University, Yangzhou 225009, China)

Abstract: Fusarium head blight (FHB) is a worldwide disease which has destructive effects on wheat production, resulting in severe yield reduction and quality deterioration, and infected wheat grains are toxic to people and animals due to accumulation of fungal toxins. However, our knowledge of mechanism underlying the host resistance is still limited due to the complexity of wheat-pathogen interactions. Recently the rise of multi-omics, including transcriptomics, proteomics and metabolomics, etc, has greatly increased our understanding of resistance mechanism controlling FHB resistance.

Here we reviewed recently published papers relevant to various omics in plant-fusarium interactions, and concluded that cell wall structure and pentose phosphate pathway (PPP) in host may associate with basal response to FHB and secondary cell wall thickening. The Krebs cycle (TCA), ethylene (ET) biosynthesis pathway, jasmonic acid (JA)/Methyl jasmonic acid (MeJA) pathway, and the early activation of salicylic acid (SA) pathway were associated with FHB resistance while photosynthesis, glycolysis (EMP), the delayed activation of SA pathway, and reactive oxygen species (ROS)/nitric oxide (NO) pathway were involved in FHB susceptibility. Ca^{2+} signaling pathway which is differently expressed after pathogen infection plays important roles in both the FHB resistance or the susceptibility, depending on signal pathway that it work with. Besides, we found that not all pathogen-related (PR) proteins are associated with FHB resistance and the resistance level varied with different PR-protein transgenic events.

By deciphering the complicated resistance mechanism of fusarium head blight (FHB)

基金项目:国家自然科学基金项目(31771772);国家重点研发计划"七大作物育种试点专项"子课题(2017YFD0100801);扬州市重点研发计划(现代农业)(YZ2016035)

通讯作者:李韬(1971-),男,甘肃会宁,教授,主要从事小麦赤霉病抗性和小麦硒生物强化的遗传基础研究;电话:15205279251,E-mail: taoli@yzu.edu.cn

第一作者:周瑶(1993-),女,江苏泰兴,硕士研究生,主要从事小麦赤霉病抗性研究;电话:17352488220,E-mail:1090362118@qq.com。

in wheat at the integral level on the basis of multi-omics, we may discover the genes involved in processes or pathways that are associated with FHB resistance, and then utilize these genes to improve FHB resistance in wheat breeding by using transgenic approaches, gene editing or marker assisted selection strategies.

Keywords: wheat; Fusarium head blight (FHB); multi-omics; resistance mechanism

A novel and nitrogen-dependent gene associates with hypersensitive reaction-like trait and its signal pathways in wheat

Xuan Shi, Lei Li, Tao Li*

(Yangzhou University, Yangzhou 225009, China)

Abstract: Wheat hypersensitive reaction-like (HRL) trait are characterized as the formation of necrotic lesions without infection by pathogens, and such genetic defects often lead to enhanced resistance to pathogen infection and constitutive expression of defense response genes. In a wheat line P7001, HRL trait was expressed at low level of nitrogen, but disappeared when sufficient N was supplied, therefore the phenotype was regulated by nitrogen and thus was nitrogen-dependent. Using bulked segregant analysis strategy together with RNA-seq technique (BSR-Seq), we identified a nitrogen-dependent HRL gene (*Ndhrl1*) and mapped it to the short arm of chromosome 2B using an F5 recombinant inbred population developed from the cross of P7001 × P216. The putative gene was located at an interval between the CAPS/dCAPS markers 7hrC9 and 7hr2dc14, and co-segregated with the dCAPS marker 7hrdc2. *Ndhrl1* was further delimited into an interval of 8.5 Mb flanked by the molecular markers 7hrdc2 and C075783 in a larger heterozygotes-derived population (F7). From the chromosomal location and the nature of N-responsiveness of the gene, *Ndhrl1* is likely to be a novel gene in wheat. Further analysis of RNA-seq data showed that several signal transduction pathways, including plant-pathogen interaction, nitrogen metabolism, zeatin biosynthesis and plant hormone, were significantly differentially expressed between HRL plants and non-HRL plants. This study is of great significance for deciphering broad-spectrum resistance mechanism and clarifying the relationships among nitrogen fertilizer, hypersensitive reaction and disease resistance.

Keywords: wheat; hypersensitive reaction-like; nitrogen-dependent gene; signal pathway

---

基金项目:国家自然科学基金项目(31600997; 31270704);扬州大学科技创新团队;江苏高校优势学科建设工程资助项目(PAPD)

通讯作者:李韬(1971-),男,甘肃会宁,教授,主要从事小麦赤霉病抗性和小麦硒生物强化的遗传基础研究;电话:15205279251,E-mail: taoli@yzu.edu.cn

第一作者:施璇,女,江苏宿迁,硕士研究生,主要从事小麦广谱抗性基因的定位研究;电话:18852717435,E-mail:1608807841@qq.com。

Investigation of the genetic diversity of *Phytophthora capsici* in China using a universal fluorescent labeling method

Xiaoren Chen[1*], Tingting Liu[1], Guanghang Qiao[2]

([1] *College of Horticulture and Plant Protection, Yangzhou University, Yangzhou 225009, China;*
[2] *Institute of Plant and Environment Protection, Beijing Academy of Agriculture and Forestry Sciences, Beijing 100097, China*)

Abstract: *Phytophthora capsici* is an important oomycete pathogen threatening the vegetable production in China, but very little is known about the population structure. The objective of the present study was to evaluate the genetic diversity of 54 *P. capsici* isolates obtained from 2010—2015 at 12 provincial locations in China. Isolates were assessed for mating type, metalaxyl resistance, and simple sequence repeat (SSR) genotype. Mating-type analyses of 54 isolates showed that both mating types were present in most (9) of the sampled production regions (12), and the mating-type frequency in the total Chinese population did not deviate significantly from a 1∶1 ratio. Responses of isolates to the fungicide metalaxyl indicated the absence of metalaxyl-resistance among the field populations. A universal fluorescent labeling method was adapted in this study to improve the efficiency of SSR genotyping. Microsatellite genotyping using 7 SSR markers was performed on 54 isolates. Genetic analyses using the program STRUCTURE indicated the existence of 2 genetic clusters within Chinese *P. capsici* collection. The same result was obtained by using the program PowerMarker. Further studies with expanded scale sampling at single locations over multiple years will be valuable to define the genetic diversity of *P. capsici* in China.

Keywords: *Phytophthora capsici*; genetic diversity; microsatellite; fluorescent labeling

---

基金项目:国家自然科学基金项目(31671971);扬州市自然科学基金项目(YZ2016121)
通讯作者:陈孝仁,男,副教授,主要从事植物病原真菌学研究;E-mail:xrchen@yzu.edu.cn。

The functional analyses of *PsAvr3c* family member from *Phytophthora* species reveal the evolution trace of these effectors

Ying Zhang, Jie Huang, Suomeng Dong**

(*Department of Plant Pathology, Nanjing Agricultural University, Nanjing 210000, China*)

Abstract: PsAvr3c is an avirulence effector from oomycete plant pathogen *Phytophthora sojae* which physically binds to GmSKRPs *in vivo*. GmSKRPs, negative immune regulators, are Serine-Lysine-Arginine rich proteins that associate with plant spliceosome and involve in the pre-mRNA splicing process. *PsAvr3c* targets to GmSKRPs to reprogram alternative splicing of more than a thousand of pre-mRNAs and consequently subvert plant immunity. However, whether this virulence mechanism is conserved in other *Phytophthora* species remains unknown. Here, we identified three *PsAvr3c* homologous effector genes from different *Phytophthora* species, including *PsAvh27b* (paralog from *P. sojae*) and *ProbiAvh89* (ortholog from *P. robiniae*), and *PparvAvh214* (ortholog from *P. parvispora*). A series of experiments demonstrate that both *PsAvh27b* and ProbiAvh89 also binds to and stabilize GmSKRP, Moreover, Analysis by qRT-PCR data indicates that alternative splicing of pre-mRNAs from 401 soybean genes, including defense related genes, is altered in PsAvh27b and ProbiAvh89 over-expressing lines compared to control plants. In summary, our data demonstrate that *PsAvr3c* family members still share conserved functions with some functional specificities.

Keywords: homologous; alternative splicing; plant immunity; evolution

---

Corresponding author: Suomeng Dong, professor, mainly engaged in blight genomics and epigenomics; E-mail: smdong@njau.edu.cn.

Identification and functional analysis of the NLP-encoding genes of the phytopathogenic oomycete *Phytophthora capsici*

Xiaoren Chen*, Guilin Sheng, Yanpeng Li, Yuping Xing

(*College of Horticulture and Plant Protection, Yangzhou University, Yangzhou 225009, China*)

Abstract: *Phytophthora capsici* is a hemibiotrophic, phytopathogenic oomycete that infects a wide range of crops, resulting in significant economic losses worldwide. By means of a diverse arsenal of secreted effector proteins, hemibiotrophic pathogens may manipulate plant cell death in order to establish a successful infection and colonization. In this study, we described the analysis of the gene family encoding necrosis- and ethylene-inducing peptide 1 (Nep1)-like proteins (NLPs) of *P. capsici*, and identified 39 real NLP genes and 26 NLP pseudogenes. Transcriptional profiling of 8 selected NLP genes indicated that they were differentially expressed during the developmental and plant infection phases of *P. capsici*. Phylogenetic analysis showed that *P. capsici* NLPs cluster closely with other oomycete NLPs, forming Type I group, which is distinct from Type 2 and Type 3 groups. Only 17 NLPs contain a fully conserved heptapeptide GHRHDWE motif that is considered to be important for phytotoxic activity, whereas the motif is degenerated in most of NLPs. Functional analysis demonstrated that Pc11951, Pc107869, Pc109174 and Pc118548 were capable of inducing cell death in *Nicotiana benthamiana* and hot pepper leaves. This study provides a first overview of *P. capsici* NLP gene family that may play important virulence roles during the plant infection.

Keywords: *Phytophthora capsici*; necrosis; NLP; virulence; agroinfiltration

---

基金项目:国家自然科学基金项目(31671971);扬州市自然科学基金项目(YZ2016121)

通讯作者:陈孝仁,男,副教授,主要从事植物病原真菌学研究;E-mail:xrchen@yzu.edu.cn。

Function analysis of SCR2 and its oomycetes orthologs

Rongkang Xing, Shuaishuai Wang, Suomeng Dong*

(College of Plant Protection, Nanjing Agricultural University, Nanjing 210095, China)

Abstract: A number of small cysteine-rich (SCR) effectors genes are highly induced at early infection stage of many oomycete plant pathogens. However, the mode of action of these SCR effectors remains largely unclear. *SCR2* from *Phytophthora infestans* encodes a novel PAMP molecule that trigger reactive oxygen burst, upregulation of defense related genes and hypersensitive response in *Solanaceous* plant. *SCR2* orthologs are wildly presented in oomycete including *Phytophthora*, *Hyaloperonospora*, *Peronospora*, *Pythium* and *Plasmopara*. Intriguingly, SCR2 from *Phytophthora* species triggers cell death on *N. benthamiana*, but SCR2 orthologs from downy mildew species cannot, suggesting sequence polymorphisms are important to SCR2 function. Indeed, mutagenesis assay identified a 60-amino acid region that is important for SCR2 function, in which two cysteines and one polymorphic residue are also identified as key sites for SCR2 function. Incubation of SCR2 protein with plant apoplast fluid indicated that SCR2s are subjected to proteolytic cleavage by plant protease, in particular serine proteases because proteolytic cleavage could be blocked by both chemical inhibitor and pathogen effector inhibitor. We also identified serine proteases from both tomato and soybean that specifically cleaved SCR2 effectors. In summary, we identified SCR2 as a PAMP molecule on *Solanaceous* plant and identified key sequence signature that is important for activity. Effector that is cleaved by host protease, as such induce plant immunity uncovers a novel perspective for plant-microbe interaction.

Keywords: oomycetes; SCR effectors; PAMP; proteolysis; immunity

Corresponding author: Suomeng Dong; E-mail: smdong@njau.edu.cn.

A single avirulent strain of *Sclerotinia sclerotiorum* co-infected by two mycoviruses shown by deep sequencing

Muhammad Rizwan Hamid, Jiatao Xie, Jiasen Cheng, Yanping Fu, Daohong Jiang[*]

(*State Key Laboratory of Agricultural Microbiology, Provincial Key Lab of Plant Pathology of Hubei Province, College of Plant Science and Technology, Huazhong Agricultural University, Wuhan 430070, China*)

Abstract: Abnormal morphology was observed in a strain named 228 of *Sclerotinia sclerotiorum*. Pathogenicity tests showed less pathogenicity on leaves of rapeseed in case of this specific strain. Bands were detected on an electrophoretic profile of double-stranded (ds) RNA preparation. The Total RNA sample was extracted and deep-sequenced by a next-generation sequencer, two virus-like assembled contigs with predicted amino acid sequences showing homologies to respective viral RNA-dependent RNA polymerases (RdRps) were found by BLAST analysis performed on NCBI. One out of two sequences showed homologies to RdRp of known mycovirus, that is; Botrytis pori RNA Virus 1 (BpRV1), and one unclassified mycovirus, named Sclerotinia sclerotiorum deltaflexivirus 2 which is closely identical to Soybean leaf-associated mycoflexivirus 1, Sclerotinia sclerotiorum deltaflexivirus 1 and Fusarium graminearum deltaflexivirus 1 respectively. Based on genome structure and phylogenetic analysis, unclassified mycovirus were taught to be member of a newly proposed family Deltaflexivirus. The presence of both mycoviruses were confirmed by sequencing RT-PCR products generated from dsRNA and Total RNA respectively. The complete genome of novel mycovirus named Sclerotinia sclerotiorum deltaflexivirus 2 (SsDFV2) was shown to be 6524 nucleotides long excluding poly (A) tail. SsDFV2 has a single large open reading frame (ORF) and encodes a putative methyltransferase-helicase-RdRp polyprotein. No evidence for a coat protein encoded by SsDFV2 was obtained. Pathogenicity tests on fresh rapeseed leaves suggested that presence of these mycoviruses decreased the pathogenicity of *Sclerotinia sclerotiorum* as compared to control (virus free strain). Horizontal viral transmission assay was performed by growing virus-free

---

Correspondence author: Daohong Jiang; E-mail: daohongjiang@mail.hzau.edu.cn.

and virus-containing strain together and it was observed that only novel mycovirus was able to transmit from infected to healthy strain, which significantly decreased the pathogenicity of infected strain. This study is to report a novel mycovirus from *Sclerotinia sclerotiorum* with strong infectivity and ability to transmit from infected to virus-free strains, which shows potential of mycovirus for biological control of fungal diseases.

Keywords: mycovirus; biological control; *Sclerotinia sclerotiorum*; deltaflexivirus; dsRNA

Detection and identification of the putatively novel badnaviruses infecting lotus (*Nelumbo nucifera* Gaertn.)

Zhen He*, Xian Liu, Chunfeng Chen, Liangjun Li

(*School of Horticulture and Plant Protection, Yangzhou University, Wenhui East Road No. 48, Yangzhou 225009, China*)

Abstract: Lotus (*Nelumbo nucifera* Gaertn.) is an economically important perennial aquatic herb plant in China. Presently, viral diseases only caused by cucumber mosaic virus (CMV) and dasheen mosaic virus (DsMV) were reported on lotus. In this study, three putatively novel badnaviruses from lotus plants were identified, tentatively named lotus badnavirus 1-3 (LBV 1-3), using next generation sequencing of siRNAs and conventional Sanger sequencing. Lotus plants infected by LBVs in the field harbor virus quasispecies, and were often mixed infected by different species of the genus *Badnavirus*. A unique recombination event was first identified in a single lotus plant. Sap inoculation showed that LBVs could actively replicate in the lotus plants with mild-mosaic or mottle symptoms. The field survey of 43 lotus plants grown in Jiangsu province showed a prevalence of 62.7% for LBVs, confirming it was widely distributed in China. However, there was no obvious connection between specific symptoms and virus infection.

Keywords: badnaviruses; mixed infection; *Nelumbo nucifera* Gaertn.; China

Correspondence author: Zhen He; E-mail: hezhen@yzu.edu.cn.

Both single amino acid on TGB1 and γb dimerization determine the pathogenicity of the *Barley stripe mosaic virus* on maize

Kun Zhang[1,2#], Yue Hu[1#], Jin Ma[1], Zhenggang Li[1], Yongliang Zhang[1], Dawei Li[1*]

([1] *State Key Laboratory of Agro-Biotechnology and Ministry of Agriculture Key Laboratory of Soil Microbiology, College of Biological Sciences, China Agricultural University, Beijing 100193, China;*
[2] *Department of Plant Pathology, School of Horticulture and Plant Protection, Yangzhou University, Yangzhou 225009, China*)

Abstract: *Barley stripe mosaic virus* (BSMV), the type member of the genus of *Hordeivius*, mainly infects barley under natural conditions, and could also infects numerous dicots and monocots by artificial inoculations. Since early studies on phenotype and host range of BSMV revealed that some BSMV strain could infects maize plant, and elicited disease symptoms. However, that systematic analysis about pathogenicity of different BSMV strains to maize plant remains elusive. In our studies, of the 260 wide-genome sequenced maize inbred lines from different stages of breeding history, 35 could be infected by ND18 (ND) strain, 16 by Xinjiang (XJ) strain, and 12 by both ND and XJ. Interestingly, the BSMV-XJ infectious clone could infect the maize plant B73, while the ND could not. Through viral genomic RNA reassortment, fragment substitution, and validation of point mutations, we showed that glutamic acid (E) at position 403 (E^{403}) of $_{XJ}$TGB1 is critical for BSMV *XJ* systemic infection of maize B73. Single aa mutation the corresponding site of NDTGB1 (G404E), generated the virus $_{ND}$BSMVG^{404E} that could systemic infection of maize. And γb deletion mutation analysis revealed that $_{ND}$BSMVG^{404E}-$\gamma b_{1\text{-}131}$ could systemic infection of maize B73, whereas $_{ND}$BSMVG^{404E}-$\gamma b_{1\text{-}127}$ could not. This revealed the tripeptide ^{128}NAI^{130} of γb is the boundary that responsible for $_{ND}$BSMVG^{404E} systemic infection on maize B73. Further research indicates that the tripeptide ^{128}NAI^{130} is necessary for γb dimerization and plays significant role in BSMV systemic infection of maize. Finally, mutant

---

Corresponding author: Dawei Li, Ph. D and Professor, China Agricultural University; Tel: +86-10-62733326, E-mail: Dawei. Li@cau. edu. cn

# These authors contributed equally to this work.

viruses inoculation analysis demonstrated that both single aa mutation of $_{ND}$TGB1 (G404E) and the tripeptide ^{128}NAI^{130} boundary of γb contribute to overcome the viral long-distance movement deficiency, and actually realized systemic infection of maize B73. Our studies provided numerous alternative maize materials for further BSMV pathogenicity studies on maize, and unambiguously pave the way for our deeper understanding of the molecular mechanisms involved in the distinct viral pathogenesis and disease outcome.

Keywords: *Barley stripe mosaic virus*; maize; pathogenesis; systemic infection; TGB1; γb

Enhancing *Rice stripe virus* resistance by silencing virus transmission-related proteins gene in *Laodelphax Striatellus*

Lifei Zhu, Jinwen Du, Yang Xu, Yunzhi Song, Changxiang Zhu*

(*Stata Key Laboratory of Crop Biology, Shandong Agricultural University, Tai'an 271018, China*)

Abstract: *Rice stripe virus* (RSV) causes devastating disease in rice fields in many East Asian countries, which is transmitted to rice by small brown planthopper (SBPH) (*Laodelphax striatellus*) through a circulative persistent manner. Virus infection and transmission requires proper interactions between host and viral factors. Interfering the interaction may be an efficient strategy to enhance RDV resistance. In this study, we artificially synthesized SPBH dsGAPDH3 and dsRACK, which were reported as transmission-related proteins to interact with viral factors. A feeding-based RNAi technique was used for down-regulating GAPDH3 and RACK expression in SPBH and the qRT-PCR assay showed that the accumulation of RSV was decreased dramatically. Moreover, the virus acquisition rate and transmission rate in SPBH reduced 22.55%, 23.67% and 31.54%, 44.44% compared with control, separately, indicating that GAPDH3 and RACK played an important role in the spread of RSV. In addition, transgenic rice plants generated through *Agrobacterium*-mediated transformation with the plant expression vector hpSP, hpGAPDH3, hpRACK, hpSP+GAPDH3 or hpSP+RACK, and the RSV resistant rates were 27.41%, 13.33%, 16.72%, 42.34% and 50.00% respectively. All these results revealed that silencing the virus gene and transmission-related proteins genes simultaneously was more efficient in preventing the spread of RSV. Southern-blotted analysis showed that different exogenous genes can integrated into the rice genome with different copy number which was not related to the disease resistance. Additionally, northern blot exhibited that the accumulation of RNA in resistant transgenic plants was significantly lower than that in the susceptible plants. And massive siRNA were detected in resistant plants, but no band was observed in non-transgenic plants. Total RNA and siRNA northern blot analysis results suggested that

This work was supported by the China National Transgenic Plant Research and Commercialization Project (Grant No. 2016ZX08001-002)

Corresponding author: Changxiang Zhu; E-mail: zhchx@sdau.edu.cn.

the resistance of transgenic plants were RNA-mediated virus resistance. Our findings reveal that the knockdown of dsGAPDH and dsRACK expression significantly prevented the spread of RSV in the bodies of SBPHs and rice.

Keywords: RNA-silencing; *Laodelphax Striatellus*; virus transmission-related protein; *rice stripe virus*; transgenic rice

Structural and functional analyses of interaction between *Turnip crinkle virus* (TCV) CP and AtSO

Chao Wu[1,2], Kunchithapadam Swaminathan[2], Sek Man Wong[1,2,3,4*]

([1] *NUS Graduate School for Integrative Sciences and Engineering, National University of Singapore, Singapore;*
[2] *Department of Biological Sciences, National University of Singapore, Singapore;*
[3] *Temasek Life Sciences Laboratory, Singapore;*
[4] *National University of Singapore Suzhou Research Institute, Suzhou Industrial Park, Jiangsu, China*)

Abstract: Sulfite oxidase (SO) is a molybdenum cofactor (Moco) containing enzymes, which reduces excess sulfite. Hibiscus SO interacts with *Hibiscus chlorotic ringspot virus* (HCRSV) by binding to the S and P domains of coat protein (CP) and triggers sulfur enhanced defense (SED). Similar phenomenon is demonstrated in *Turnip crinkle virus* (TCV). How does SO interact with the virus at structural level, and by what means they regulate the sulfur pathway? To answer these questions, we aim to find out the crucial residues by analyzing the crystal structure of AtSO and TCV CP. The binding regions from both proteins were narrowed down to the moco domain of AtSO and 264-281 aa of TCV CP. A "Glycine-rich" linker was added in between the protein and the peptide sequence to be expressed together. Several crystals formed, and diffracted with 3Å resolution. However, no signal of the peptide was detected. Furthermore, TCV CP Trp274 was the key residue for the binding. No binding between TCV CP and AtSO was observed when Tryptophane (W) was mutated to Alanine (A). Furthermore, plants inoculated with the mutated TCV RNA, the expression of AtSO and SED-related genes was not different from the mock plants. Interestingly, AtSO was found to compete with AGO1 to bind to TCV CP through Trp274. These results suggested that the AtSO is a host anti-viral factor that interferes viral silencing suppressor (VSR) functions of TCV CP. In conclusion, the key residue for TCV CP and AtSO binding is identified as Trp274. AtSO competes with AGO1 to bind with TCV CP. These novel findings explained the important role of AtSO as an anti-viral factor in SED in *A. thaliana*.

Corresponding author: Sek Man Wong; E-mail: dbswsm@nus. edu. sg.

The comparative study of multiple small RNA combinations mediated virus-resistance to PVY^{O} and PVY^{N}

Ru Yu, Caixia Chen, Yunzhi Song, Hongmei Liu, Changxiang Zhu*

(*Stata Key Laboratory of Crop Biology, Shandong Agricultural University, Tai'an 271018, China*)

Abstract: *Potato virus Y necrotic strain* (PVY^{N}) and *Potato virus Y common strain* (PVY^{O}) were two major lines of *Potato virus Y* (PVY), causing systemic necrosis and systemic mottled symptoms in tobacco, respectively. The siRNA or miRNA-mediated gene silencing can inhibit the virus replication in plants. However, under natural conditions, the virus is highly susceptible to mutations thus affecting the cleavage efficiency of the sRNAs to the viral genome. Therefore a single sRNA may not be sufficient to give the plant a high degree of resistance to the virus. In this study, two regions of PVY^{O} replicase gene (*NIb*) were used as target sequences to design siRNA and miRNA. Eight plant expression vectors were constructed as psiR1, psiR2, pmiR1, pmiR2, psiR1siR2, pmiR1miR2, psiR1miR2 and psiR2miR1, meanwhile luciferase (*LUC*) reporter and NIb fusion expression vectors were constructed as 35S:PVY^{O}-NIb-LUC, 35S:PVY^{N}-NIb-LUC, which were respectively co-infected *Nicotiana benthamiana* leaves with eight sRNAs expression vectors by *Agrobacterium* transient expression system. Our results show that *NIb* of PVY^{O} and PVY^{N} is successfully cleaved by those sRNAs, and the effect of a plasmid carried siRNA and miRNA is the strongest. Similar findings were obtained in analysis of resistance to PVY^{O} and PVY^{N} infection in *Nicotiana tabacum* transgenic plants expressing PVY^{O}-NIb sRNAs. We speculate this phenomenon might be consistent with a method the siRNA-pathway and miRNA-pathway are complementary and interactive in gene silencing. We also show that sRNA activity was affected by the mismatched bases between the sRNA and the target gene, and sRNAs-mediated viral resistance could be inherited in plant progeny steadily. These findings confirm that multiplexing siRNA and miRNA appears to provide a promising resistant approach for virus infection.

Keywords: *Potato virus Y*; NIb; amiRNA; siRNA; RNAi

This work was supported by National Science Foundation of China(31272113)

Corresponding author: Changxiang Zhu; E-mail: zhchx@sdau.edu.cn.

The role of plant ribosome-inactivating proteins in defense against pathogen attack

Yangkai Zhou, Yanping Che, Shichang Cheng, Zheng Liu, Feng Zhu*

(*College of Horticulture and Plant Protection, Yangzhou University, Yangzhou 225009, China*)

Abstract: Ribosome-inactivating proteins (RIPs) are toxic *N*-glycosidases that depurinate the universally conserved α-sarcin loop of large eukaryotic and prokaryotic rRNAs, thereby arresting protein synthesis during translation. RIPs are widely found in various plant species and within different tissues. RIPs in plants have been linked to defense by antibacterial, antiviral, antifungal, and anti-insecticidal properties demonstrated *in vitro* and in transgenic plants. However, the mechanism of these effects is still not completely clear. There are a number of recent reviews of RIPs. However, there are no reviews on the biological functions of RIPs in defense against pathogens and insect pests. Therefore, we focused on the role of plant RIPs in defense against pathogens and insect pest attacks. The role of plant RIPs in defense against pathogens and insects is just beginning to be understood. Future research employing transgenic technology approaches to study the mechanisms of RIPs will undoubtedly lead to a better understanding of the role of plant RIPs in defense against pathogens and insects. These studies will help to reveal important points of genetic control that can be useful to engineer crops tolerant to biotic stress.

This work was supported by the National Natural Science Foundation of China (31500209), Natural Science Foundation of the Higher Education Institutions of Jiangsu Province of China (15KJB210007) and Natural Science Foundation of Yangzhou (YZ2015106)

Corresponding author: Feng Zhu; E-mail: zhufeng@yzu. edu. cn.

尖孢镰刀菌不同专化型过氧化氢酶活性差异分析

王亚军，洪坤奇，李成成，文才艺*

（河南农业大学植物保护学院，郑州 450002）

摘要：尖孢镰刀菌是一种分布广泛的土壤丝状真菌，可引起作物枯萎病并具有高度的寄主专化性，但目前有关尖孢镰刀菌的致病性分化及其致病机理并不清楚。已有研究表明活性氧（ROS）及过氧化氢酶（CAT）在丝状真菌生长发育和侵染过程中可发挥重要作用。为探析CAT在尖孢镰刀菌不同专化型致病过程中的功能，在初步的研究中我们以不同专化型菌株为材料分析了过氧化氢酶活性水平。通过 qRT-PCR 分析和过氧化氢酶酶谱凝胶试验操作，发现不同专化型尖孢镰刀菌中基因 *catalase*-1 和 *catalase* 表达水平以及过氧化氢酶的活性酶谱存在明显差异。其中，在苦瓜、甜瓜专化型尖孢镰刀菌中基因 *catalase* 的表达水平较低；而在西瓜、丝瓜、黄瓜专化型中该基因表达水平较高（图 1-A）。这一结果与过氧化氢酶酶谱凝胶试验一致（在苦瓜、甜瓜专化型尖孢镰刀菌中过氧化氢酶活性较低，而在西瓜、丝瓜、黄瓜专化型中过氧化氢酶活性较高）。同时，在黄瓜、丝瓜专化型尖孢镰刀菌中，过氧化氢酶活性条带更为弥散，表明在这些菌株中 CAT 蛋白上可能发生了修饰（图 1-B）。外源氧胁迫（H_2O_2）的敏感性分析实验显示，同苦瓜、甜瓜专化型尖孢镰刀菌相比，西瓜、丝瓜、黄瓜专化型菌株具有更强的 H_2O_2 抗性（图 2）。这一现象与不同专化型菌株中基因 *catalase* 表达水平和过氧化氢酶活性的水平差异一致，表明 *catalase* 基因可能是苦瓜、甜瓜、西瓜、丝瓜、黄瓜专化型尖孢镰刀菌中主要的过氧化氢酶基因。本研究的实验结果，将为进一步研究 CAT 在尖孢镰刀菌不同专化型菌株生长发育及致病过程中的功能提供基础。

关键词：尖孢镰刀菌；专化型；过氧化氢酶；外源氧胁迫

基金项目：公益性行业（农业）科研专项（201503110）

通讯作者：文才艺；E-mail：wencaiyi1965@163.com

第一作者：王亚军；E-mail：13526490191@163.com。

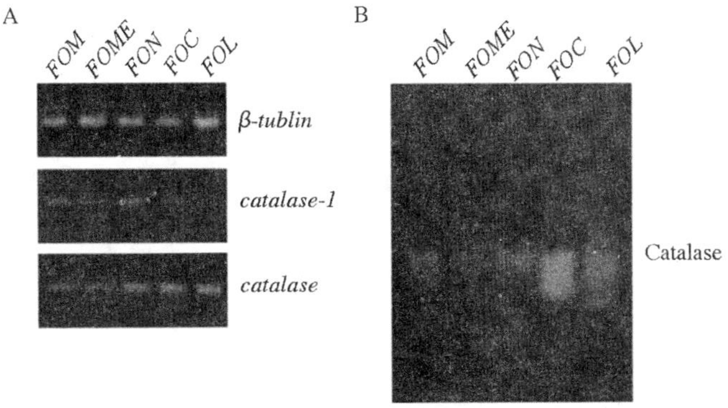

图 1　不同专化型尖孢镰刀菌中基因 *catalase-1*、*catalase* 表达水平分析和过氧化氢酶活性检测结果分析
A:qRT-PCR 分析基因表达水平。B:过氧化氢酶酶谱凝胶试验分析。尖孢镰刀菌苦瓜专化型,*FOM*;尖孢镰刀菌甜瓜专化型,*FOME*;尖孢镰刀菌西瓜专化型,*FON*;尖孢镰刀菌黄瓜专化型,*FOC*;尖孢镰刀菌丝瓜专化型,*FOL*。

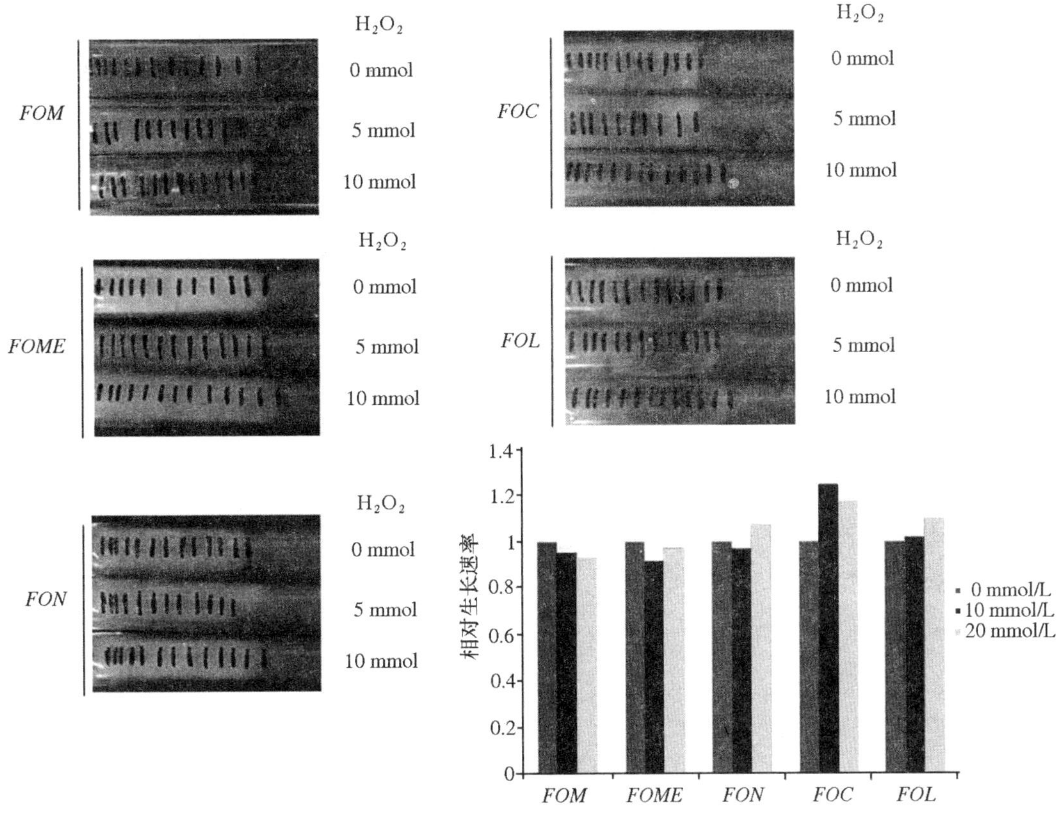

图 2　不同专化型尖孢镰刀菌在 0、10、20 mmol/L H_2O_2 培养条件下菌丝的生长状况
将菌株分别接种在含 0、10、20 mmol/L H_2O_2 的 30 cm 竞争性生长管培养基中在 25℃ 光照条件下培养。每天在同一时刻标记菌丝的生长前沿,计算菌株的平均生长速率和相对生长速率。

A-to-I RNA 编辑和可变剪接对禾谷镰刀菌 *FgBUD14* 基因的协同调控作用

梁洁[1]，刘慧泉[1]，江聪[1]，许金荣[1,2]

（[1]西农-普度大学联合研究中心/旱区作物逆境生物学国家重点实验室/
西北农林科技大学植物保护学院，杨凌 712100；
[2]美国普渡大学植物及植物病理系，印第安纳州 IN47907）

摘要：禾谷镰刀菌（*Fusarium graminearum*）作为优势种引起的小麦赤霉病（Fusarium head blight）是一类世界性麦类流行病害，它不仅会造成小麦产量降低，还会在染病小麦籽粒中积累 DON、ZEA 等真菌毒素，威胁人畜健康，造成严重的食品安全问题。禾谷镰刀菌的子囊孢子是引起小麦赤霉病的初侵染源，因此，研究禾谷镰刀菌有性发育调控对小麦赤霉病防控具有重要意义。实验室前期工作发现禾谷镰刀菌有性时期特异发生 A-to-I RNA 编辑现象。有意思的是，我们发现禾谷镰刀菌中与酵母出芽相关基因 *BUD14* 同源的基因 *FgBUD14* 的编码区内存在一个提前终止密码子 TAG，该密码子在有性发育阶段既可以通过可变剪接去除，也可以被 A-to-I RNA 编辑修复，从而恢复编码完整蛋白的能力。为了探究可变剪接和A-to-I RNA 编辑对该基因的协同调控作用，首先对 FgBUD14 进行了敲除分析，获得了 *Fgbud14* 缺失突变体。营养生长方面，*Fgbud14* 气生菌丝形态正常，但生长速度相比野生型 PH-1 降低了 16.3%；无性繁殖方面，Fgbud14 突变体产分生孢子的数量仅为野生型的40.3%，且分生孢子变短，分隔减少，萌发推迟；有性生殖方面，*Fgbud14* 突变体子囊数量减少，子囊发育滞缓，部分子囊孢子畸形且子囊孢子丧失了喷发能力。此外，*Fgbud14* 突变体的致病性也存在缺陷。互补菌株 *Fgbud14*/*FgBUD14* 的生长、侵染及有性发育均回复到野生型的水平。综上所述，*FgBud14* 在禾谷镰刀菌营养生长、无性繁殖、有性发育和致病性方面均具有重要调控作用。目前正在验证 A-to-I RNA 编辑和可变剪接在 *FgBUD14* 功能发挥中的作用。

关键词：禾谷镰刀菌；*FgBUD14*；RNA 编辑；可变剪接

蓖麻葡萄孢(*Amphobotrys ricini*)群体多样性研究

张蕊,吴明德,杨龙,李国庆,张静*

(农业微生物国家重点实验室,华中农业大学植物科学技术学院,武汉 430070)

摘要:蓖麻葡萄孢(*Amphobotrys ricini*)是一种葡萄孢属类似真菌。主要侵染大戟科植物,包括蓖麻、铁苋菜、一品红、地锦、虎刺梅和佛肚树等,危害植物组织造成茎腐、枯斑、花腐、果腐等症状。我国有蓖麻葡萄孢记录,但没有对其群体多样性开展研究。本研究从铁苋菜(*Acalypha australis*,大戟科)和蓖麻上分别分离得到 *A. ricini* 菌株 160 株和 3 株。对其培养特征、致病力和遗传多样性进行了分析。在培养特性方面,根据其在 PDA 培养基上的培养特征,可将 160 株 *A. ricini* 菌株分为五类:Ⅰ型菌落边缘不整齐,产黄色色素,24 株菌株属于这种类型,占 15%;Ⅱ型菌落表面覆盖大量白色绒毛状菌丝,50 株菌株属于这种类型,占 31%;Ⅲ型菌落边缘不整齐,菌落表面形成大量小菌核,41 株菌株属于这种类型,占 25.4%;Ⅳ型菌落边缘不整齐,菌落表面形成大量棕色分生孢子簇,无菌核或形成少量菌核,20 株菌株属于这种类型,占 12.7%;Ⅴ型菌落表面形成大量菌核,菌核表面覆盖白色绒毛,25 株菌株属于这种类型,占 15.4%。在致病力方面,对 160 株病菌对离体蓖麻叶片进行致病力测定(20℃,保湿,72 h)。根据病斑直径将其分为三类:132 株菌株(占供试菌株的 82.2%)的致病力强,病斑直径为 20～53 mm;23 株菌株(占供试菌株的 14.4%)致病力中等,病斑直径为 10～20 mm;5 株菌株(占供试菌株的 3.4%)致病力较弱,病斑直径小于 8 mm。进一步分析表明:菌株的菌落形态和致病力强弱没有关系。采用 RAPD 技术对其中 92 株菌株进行了遗传多样性分析,9 条 10-bp 寡核苷酸引物共检测到 54 个位点,多态位点 30 个,多态百分率 55.6%,表明 *A. ricini* 病菌具有丰富的遗传多样性。利用 NTSYSpc2.1 软件对多态性条带进行聚类分析,结果显示 92 株菌株遗传相似系数变化幅度在 0.56～0.95 之间,在遗传相似系数为 0.79 的水平上,可明显分为 6 个亚群(A、B、C、D、E、F),其中 A 亚群有 84 株,占 91%;B 亚群有 2 株,占 2.2%;C 亚群有 2 株,占 2.2%;D 亚群有 1 株,占 1%;E 亚群有 2 株,占 2.2%;F 亚群有 1 株,占 1%。进一步分析表明:培养形态相同的菌株并没有完全聚在一起,地理来源相同的菌株聚在一起,说明该菌遗传多样性与形态多样性没有直接相关性,和地理来源有直接相关性。这些结果有助于认识蓖麻葡萄孢的变异以及种类鉴定。

关键词:蓖麻灰霉病;培养形态;致病力;RAPD 标记;遗传多样性

基金项目:国家自然科学基金(31301615)

通讯作者:张静,副教授,主要从事植物病原真菌研究;E-mail: zhangjing1007@mail.hzau.edu.cn

第一作者:张蕊,硕士研究生;E-mail:2567264059@qq.com。

禾谷镰刀菌中 RNA 编辑对 *GCN5* 基因的调控功能

陈乐[1]，江聪[1]，刘慧泉[1]，许金荣[1,2]

([1]西农-普度大学联合研究中心/旱区作物逆境生物学国家重点实验室/
西北农林科技大学植物保护学院，杨凌 712100；
[2]美国普渡大学植物及植物病理系，印第安纳州 IN47907)

摘要：小麦赤霉病是一种世界范围内广泛流行的农作物病害，除了造成小麦产量降低，还会在感病小麦中大量积累真菌毒素 DON，引发严重的食品安全问题。禾谷镰刀菌是小麦赤霉病的主要致病菌，其有性子囊孢子作为重要的初侵染源，在小麦赤霉病的发生过程中起到了至关重要的作用。本实验室前期研究发现，禾谷镰刀菌有性发育阶段存在 A-to-I RNA 编辑现象，A-to-I RNA 编辑导致有性阶段表达的半数以上基因编码的蛋白产物发生改变。其中，组蛋白乙酰转移酶基因 *GCN5* 表达的 22%转录本的终止密码子 TAG 在有性发育阶段被 RNA 编辑后变成 TGG，出现翻译通读，因此，*GCN5* 在有性发育阶段可同时编码一个正常大小的蛋白和一个延长了 50 氨基酸的长蛋白。由于组蛋白乙酰转移酶介导的组蛋白乙酰化是一种重要的表观修饰机制，可参与调控基因的转录和沉默、染色质凝聚、DNA 复制、胁迫应答等多种细胞生命过程，因此，有性发育阶段特异出现的 Gcn5 长蛋白很可能具有一些特异的调控功能。为了探究 A-to-I RNA 编辑及长蛋白的功能，我们对禾谷镰刀菌的 *GCN5* 基因进行了敲除分析，发现敲除突变体菌落生长滞缓、气生菌丝少、不产孢、致病能力完全丧失。此外，突变体有性发育发生严重缺陷，不能产生子囊壳。外源引入野生型 *GCN5* 片段到突变体中可以完全恢复 *Δgcn5* 突变体的表型；单独引入不能通读的 *GCN5* 片段后，互补菌株在营养生长、产孢和致病力等方面均能恢复到野生型水平，有性发育部分恢复，能产生子囊壳，但子囊和子囊孢子发育以及子囊孢子喷发存在缺陷；单独引入编辑后的 *GCN5* 片段后，互补菌株有性阶段表型不能恢复，但无性阶段表型部分恢复。综上所述，*GCN5* 基因终止密码子上的 A-to-I RNA 编辑直接影响了其功能，参与调控禾谷镰刀菌的有性发育。该研究有助于揭示 RNA 编辑与组蛋白修饰这两大表观修饰机制之间的联系，为有效开展小麦赤霉病防控工作提供理论参考。

关键词：禾谷镰刀菌；组蛋白；乙酰化修饰；A-to-I RNA 编辑

小麦赤霉病扩展抗性和籽粒 DON 积累在不同接种方法下的差异分析

龚璇，李长成，骆孟，李磊，李韬*

（扬州大学，扬州 225009）

摘要：小麦赤霉病是由禾谷镰刀菌引起的一种世界性病害，不仅能够引起小麦减产和品质下降，而且受赤霉病感染的籽粒中积累的脱氧雪腐镰刀菌烯醇（DON）等毒素严重影响人畜健康。为了解小麦赤霉病抗性与籽粒毒素积累之间的关系，以及诱导 DON 毒素产生的组织，本研究以遗传基础广泛且抗性 QTL 组成已知的 30 份小麦品种（系）为材料，在扬花期利用单花滴注法（SFI）和基部穗轴注射法（BRI）接种鉴定赤霉病抗性，成熟后收获籽粒，利用酶联免疫法（ELISA）分别测定两种接种方法下的籽粒 DON 含量。结果表明，两种接种方法下，苏麦 3 号的感病小穗率（PSS）与宁 7840、Tokai66、NyuBai、白三月黄、黄蚕豆相似，但是其 DON 含量显著高于上述品种（系），可能与苏麦 3 号在 2DS 上携带的感病 QTL 与毒素积累有关；在 BRI 接种方法下携带 Fhb1 的抗病材料中未观察到感病小穗，但在其籽粒中仍然能够检测到较高浓度的 DON，表明穗轴是诱导毒素产生的主要组织，携带 Fhb1 的材料 DON 浓度最高可达 724.77 ppb（黄方柱），其次为苏麦 3 号（625.24 ppb）。以两种接种方法下 PSS 和籽粒 DON 含量均最低、抗性最好的品种（系）Tokai66（PSS 介于 0～0.05 之间，DON 浓度介于 126.04～146.62 ppb 之间）为对照（标准），采用基于马氏距离的综合评价方法，分析各个品种（系）与对照之间的差异，并筛选到白三月黄、宁 7840、NyuBai、黄蚕豆、Tawanhsiaomai 以及望水白的总体抗性优于苏麦 3 号，这些品种（系）在培育兼抗赤霉病扩展和抗毒素积累的品种中可能更具有利用价值。

关键词：小麦；赤霉病；接种方法；DON

基金项目：国家自然科学基金项目（31771772）；国家重点研发计划“七大作物育种试点专项”子课题（2017YFD0100801）；扬州市重点研发计划（现代农业）（YZ2016035）；江苏省高校优势学科建设工程资助项目（PAPD）

通讯作者：李韬，博士，教授，主要从事小麦赤霉病抗性和小麦硒生物强化的遗传基础研究；E-mail：taoli@yzu.edu.cn

第一作者：龚璇，硕士研究生，主要从事小麦赤霉病抗性的遗传基础研究；E-mail：703475819@qq.com。

小麦与赤霉菌互作中的蛋白质组学分析

刘家俊，李磊，李韬*

（扬州大学，扬州 225009）

摘要：小麦赤霉病是一种全球性的病害，既会导致减产，又能在籽粒中富集呕吐毒素。尽管已报道的相关研究尝试从转录组、蛋白组以及代谢组角度解析小麦赤霉病与病原菌互作机制，但其抗病分子机理仍然未知，并且这些研究都是利用单一的抗、感基因型，这在一定程度上难以从全局的角度解释抗感差异以及筛选抗病蛋白或感病因子。我们选择了 10 个抗性组成已知的抗病品种和 6 个感病品种，采用抗、感穗轴混池的方法，结合 Label-free 定量蛋白技术，从赤霉菌与小麦互作的全局视角分析了差异表达蛋白，排除了单一抗/感品种表达特异性。我们构建了 4 个穗轴蛋白质混池（抗病接种池、抗病对照池、感病接种池和感病对照池），分析结果表明，抗病对照池与抗病接种池共有 237 个差异蛋白，感病对照与感病接种池有 238 个，抗病对照池与感病对照池有 234 个，抗病接种池与感病接种池有 206 个。另外，前两者的共同差异蛋白计 38 个，后两者的共同差异蛋白共 39 个。通过比较分析发现翻译起始因子 eIF4E-1 和 G3 胚胎发育晚期丰富蛋白可能与抗病关联，而紫色酸性磷酸酶和活性氧则可能为感病因子。同时，我们根据 10 个抗病品种已报道的 QTL 构建了一致性图谱，并将蛋白映射到对应的 QTL 区间。该项研究对进一步解析赤霉病抗性机制、筛选和克隆抗赤霉病候选基因以及改良赤霉病抗性都具有一定的意义。

关键词：小麦；赤霉病；蛋白质组；QTL

基金项目：国家自然科学基金项目(31771772)；国家重点研发计划“七大作物育种试点专项”子课题(2017YFD0100801)；扬州市重点研发计划(现代农业)(YZ2016035)；江苏省高校优势学科建设工程资助项目(PAPD)

通讯作者：李韬，博士，教授，主要从事小麦赤霉病抗性和小麦硒生物强化的遗传基础研究；E-mail：taoli@yzu. edu. cn

第一作者：刘家俊，硕士研究生，主要从事小麦赤霉病抗性的遗传基础研究；E-mail：703475819@qq. com。

草莓胶孢炭疽菌候选效应子的筛选

何成勇[1,2],张丽勍[2],宋丽丽[1,2],高清华[2],段可[1,2*]

([1]上海海洋大学食品学院,上海 201306;
[2]上海市农业科学院林木果树研究所,上海 201403)

摘要:草莓胶孢炭疽菌(*Colletotrichum gloeosporioides*)是草莓炭疽病的主要致病菌之一,该病害在我国南方各草莓种植区普遍发生,造成严重的经济损失。病原真菌通过分泌一系列的效应子来调控寄主植物的抗性。本研究以实验室前期所构建的 RNA-seq 数据库为基础,筛选获得了 52 个在不同侵染阶段上调表达的候选效应子。本研究选取了其中 11 个候选效应子进行了系统鉴定。将候选效应子的全长序列及去信号肽序列分别构建至瞬时表达载体,并进行相应的亚细胞定位分析及细胞凋亡诱导或抑制筛选分析。结果表明,11 个候选效应子中,2 个候选效应子全长及去信号肽蛋白为核膜共定位,其余皆为膜定位;其中,6 候选个效应子全长及去信号肽蛋白可抑制 Bax 诱导的细胞凋亡。目前,6 个抑制细胞凋亡的候选效应子的突变体已构建完成,后续将对 6 个候选效应子的基因功能进行深入的研究,以期明确胶孢炭疽菌的致病机理。

关键词:炭疽病;胶孢炭疽菌;效应子

基金项目:国家自然科学基金(31501592);上海市科委农业科技重点攻关项目(16391901400);上海市农委种业项目(沪农种字 2017 第(2-1)号)

通讯作者:段可,研究员,主要从事草莓抗病机理研究;E-mail:kduan936@126.com
第一作者:何成勇,硕士研究生。

草莓炭疽病病原鉴定及生物学特性分析

宋丽丽[1,2]，张丽勍[2]，何成勇[1,2]，高清华[2]，段可[1,2*]

([1] 上海海洋大学食品学院，上海 201306；
[2] 上海市农业科学院林木果树研究所，上海 201403)

摘要：草莓炭疽病是严重威胁草莓生产的重要病害之一，发生普遍，分布广泛。对取自上海市和安徽省合肥市的草莓炭疽病病样进行分离纯化，应用柯赫氏法则测定了纯培养物的致病性。根据形态学特性和多基因联合建树法对病原菌进行鉴定，并进行生物学特性研究。从病样上共分离纯化到6个致病菌株，其中：来源于上海市3个菌株，合肥市3个菌株。分离自合肥市的菌株GQHAH4在PDA培养基上的菌落为白色，其余菌株均为中心深灰色，边缘灰白整齐。分生孢子整体均呈椭圆形或一端略尖。通过基于ACT、APMAT、CAL、GADPH和ITS的多基因系统进化分析结果表明，这6个菌株与已鉴定为*Colletotrichum fructicola*的菌株CBS 130416、ICMP 10642、LC2924和LF131等归在1个分支。生物学特性研究结果表明，这6个菌株最适生长温度为25～30℃，最佳pH为6～7，且能利用多种碳、氮源，但不同来源菌株生长的最优碳源、氮源有一定差异。分离自上海市的GQH26和GQH131菌株的产孢量显著高于其他菌株。本研究为明确草莓炭疽病的病原，进一步探索各地区草莓炭疽病的发病规律，预测预报及综合防治提供参考依据。

关键词：草莓；炭疽病；生物学特性；*Colletotrichum fructicola*

基金项目：国家自然科学基金（31501592）；上海市科委农业科技重点攻关项目（16391901400）；上海市农委种业项目（沪农种字2017第(2-1)号）

通讯作者：段可，研究员，主要从事草莓抗病机理研究；E-mail：kduan936@126.com
第一作者：宋丽丽，硕士研究生。

稻曲病菌厚垣孢子越冬能力的研究

雍明丽[1,2]，刘亦佳[1]，刘永锋[2]，胡东维[1*]

([1]浙江大学农业与生物技术学院生物所，杭州 310058；
[2]江苏省农业科学院植物保护研究所，南京 210014)

摘要：对稻曲病菌厚垣孢子萌发率测定的方法进行改良后，发现光照和变温处理能明显促进厚垣孢子的萌发。休眠的厚垣孢子萌发需要 5～30 d 时间，因此传统的萌发测定方法严重低估了厚垣孢子的萌发能力。进一步对不同储存时间的厚垣孢子进行萌发试验，结果表明，在室内干燥和低温环境条件下的厚垣孢子可长时间保持萌发活力，其中冷藏保存 21 个月的墨黑色厚垣孢子累计萌发率可达到 72.3%。越冬后稻田地表厚垣孢子萌发率高于土壤中厚垣孢子。田间地表稻曲球上厚垣孢子在来年 3～7 月仍然能够萌发产生分生孢子。模拟田间淹水试验表明，厚垣孢子易被土壤中微生物降解而无法萌发。以上结果表明，田间及室内保存稻曲球上厚垣孢子在干燥条件下能够安全越冬，且数量远远大于以前研究所报道的比例。但厚垣孢子只能在相对干燥环境下越冬，稻田灌水不利于厚垣孢子越冬。因此，稻曲病菌厚垣孢子在病害流行中的作用仍然有待进一步的研究。

关键词：稻曲病菌；厚垣孢子；越冬；萌发

基金项目：国家自然科学基金(31271999)

通讯作者：胡东维，研究员；E-mail：hudw@zju.edu.cn。

稻曲菌(*Villosiclava virens*)有性生殖阶段分子调控机制初析

于俊杰[1],俞咪娜[1],聂亚锋[1],孙文献[2],尹小乐[1],赵洁[1],
王亚会[1],丁慧[1],齐中强[1],杜 艳[1],Li Huang[3],刘永峰[1*]

([1]江苏省农业科学院植物保护研究所,南京 210014;
[2]中国农业大学植物保护学院,北京 100193;
[3]蒙大拿州立大学植物科学与植物病理学系,波兹曼 59717-3150)

摘要:近年来稻曲病在我国各稻区发生呈总体上升趋势,已成为水稻的重要病害之一。稻曲菌为异宗配合真菌,其有性生殖产生的子囊孢子为稻曲病的重要初侵染源,然而其有性生殖的分子调控机制尚不清楚。本研究以菌株 UV-8b 基因组测序数据为基础,比较分析了稻曲菌在子实体发育和产分生孢子阶段转录组差异。结果表明,有 488 个基因在子实体形成阶段上调表达,通过 KEGG 和 GO 数据库富集分析发现,这些基因涉及交配型基因(*MAT*1-1-2、*MAT*1-1-3)、信息素合成和分泌相关基因(KDB16470.1、KDB16030.1、KDB18386.1、KDB18574.1)、信号传导途径(KDB18934.1、KDB17304.1、KDB15053.1 和 KDB11415.1 等)、转录调控因子(KDB10897.1、KDB11819.1 和 KDB14847.1 等)和有丝分裂相关调控基因(*Ndt80*、*Spo11* 和 *Mei3* 等);此外,有 342 个基因在分生孢子形成阶段上调表达,涉及生物合成和能量代谢相关基因、分生孢子形成调控基因和病原菌侵染相关基因等。本研究为深入揭示稻曲菌有性生殖机制的研究奠定了基础。

关键词:稻曲菌;有性生殖;转录组比较

基金项目:国家自然科学基金(31571961);江苏省农业自主创新资金项目(CX(12)5005, CX(15)1054)
通讯作者:刘永锋,研究员,主要从事水稻病害及其生防研究;E-mail:liuyf@jaas.ac.cn
第一作者:于俊杰,副研究员,主要从事水稻病原真菌病理学研究;E-mail:jjyu@jaas.ac.cn。

油酸激活型转录因子 MoOaf22 参与调控稻瘟病菌的营养生长、细胞壁完整性和对外界胁迫的应答的研究

齐中强，杜艳，沈乐融，余振仙，于俊杰，俞咪娜，宋天巧，张荣胜，刘永锋*

（江苏省农业科学院植物保护研究所，南京 210014）

摘要：稻瘟病菌（*Magnaporthe oryzae*）引起的稻瘟病是生产上一种最具破坏性的真菌病害，严重威胁我国的水稻安全生产。深入了解稻瘟病菌致病过程的分子机制尤为重要。为了研究油酸激活型转录因子 MoOaf22 编码基因 MoOAF22（MGG_03413）在稻瘟病菌生长发育及致病过程中的功能，利用基因敲除方法将其进行了缺失突变并对突变体进行了一系列的生物学表型分析。结果表明，该基因缺失突变体在完全培养基（CM）上营养生长加快，但在基本培养基（MM）上生长变慢；同时该突变体原生质体释放速度变慢；对荧光增白剂（CFW）和刚果红（Congo Red）耐受性增强，对十二烷基磺酸钠（SDS）、氯化钠（NaCl）和山梨醇（Sorbitol）敏感，但该突变体在产孢量、附着胞形成及致病性等方面较野生型没有明显变化。上述结果说明 MoOaf22 在稻瘟病菌的营养生长、细胞壁完整性和对外界的胁迫应答过程中具有重要的调控作用。

基金项目：国家自然科学基金青年项目（31401697）；江苏省自然科学青年基金项目（BK20140749）；江苏省农业科技自主创新基金项目[CX(15)1054]

通讯作者：刘永锋，研究员，主要从事植物病理学研究；E-mail：Liuyf@jaas.ac.cn

第一作者：齐中强（1985- ），助理研究员，主要从事植物病理学研究；E-mail：qizhongqiang2006@126.com。

稻瘟菌中一个含 KH 结构域致病蛋白的作用机制研究

潘嵩，周威，阴长发，戚琳璐，杨俊*，彭友良

（中国农业大学植物病理系，北京 100193）

摘要：由稻瘟菌引发的稻瘟病是在世界各水稻产区广泛发生且危害严重的真菌病害，深入解析稻瘟菌的致病机理将有助于发现控制稻瘟病的新途径。本课题组前期鉴定了一个编码含多 KH 结构域蛋白的新基因 *PCG*6，发现该基因是稻瘟菌致病和无性发育所必需的。为了阐明 Pcg6 的分子功能，我们首先观测了 Pcg6-GFP 的亚细胞定位，发现其主要定位于细胞质，在营养菌丝、分生孢子、附着胞和侵染菌丝等阶段均有表达，且在分生孢子和附着胞形成阶段表达量较高。共定位结果显示 Pcg6-GFP 与 P-body 的标记蛋白 Dcp2-RFP 在细胞质中共定位。接下来我们利用免疫共沉淀和质谱技术筛选了 Pcg6 的互作蛋白，发现 Pcg6 与 RNA 降解途径相关蛋白（如 MoNmd3 和 MoCaf1 等）免疫共沉淀，且 MoNmd3 与 MoCaf1 也免疫共沉淀。进一步发现 Pcg6-GFP 与 MoNmd3-RFP 和 MoCaf1-RFP 均能共定位，暗示这 3 个蛋白处在复合体 P-body 中。酵母双杂交结果表明 MoNmd3 和 MoCaf1 与 Pcg6 直接互作，且二者均与 $Pcg6^{418\text{-}488}$ 互作。为了解析 Pcg6 对 MoNmd3 和 MoCaf1 所介导的 RNA 降解过程可能产生的影响，我们发现在 *PCG6* 敲除体中 MoNmd3 和 MoCaf1 的蛋白含量均上调。进一步在野生菌 P131 种超表达这两个蛋白，发现对应的转化子均出现了产孢能力明显下降，推测 *PCG6* 敲除体产孢缺陷是部分由于 MoNmd3 和 MoCaf1 超表达所引起。比较转录组分析发现在 *PCG6* 敲除体中有 1000 多个基因出现下调表达，其中与附着胞形成相关的两个基因 *MPG1* 与 *PTH11* 均下调了 10 倍左右，与附着胞介导的侵染相关的基因 *MoVRF1* 与 *MoBUF1* 分别下调了 5 倍以上。同时我们也功能分析了受 Pcg6 正调控的一个编码 Sit4 磷酸化相关蛋白的新基因 *MoSIPP1*，该基因的敲除体在无性发育和侵染过程均出现明显缺陷。与之相反的是，在 PCG6 敲除体中有 1000 多个基因出现上调表达，如 G 蛋白调节因子基因 *MoRGS1* 上调了 2 倍左右。在野生菌 P131 中超表达 *MoRGS1* 导致菌株产孢能力明显下降。这些结果表明 Pcg6 可通过调控不同靶标基因的转录水平来体现其生物学功能。

基金项目：国家自然科学基金项目（31371885）；中国科协"青年人才托举工程"扶持人才项目（2015QNRC001）

通讯作者：杨俊，副教授，主要从事植物病原物致病机理、植物与病原物互作组学的研究；E-mail：yangj@cau.edu.cn。

稻瘟菌外泌蛋白的糖基化修饰对其功能影响的研究

刘宁，黄曼娜，曹淼，卢训莉*，彭友良

（中国农业大学植物保护学院植物病理学系，北京 100193）

摘要：水稻是我国的第一大粮食作物，由稻瘟菌侵染水稻引起的稻瘟病，是水稻的三大病害之一，严重威胁着我国水稻的安全生产。研究稻瘟菌的致病机理，将为稻瘟病的防治提供理论基础，对生产上有效防控稻瘟病的流行具有指导作用。稻瘟菌在侵染水稻时，其外泌蛋白被分泌及转运到水稻细胞的胞内或质外体中，对于稻瘟菌的成功侵入发挥着关键作用。在外泌蛋白中，有一大类属于糖基化蛋白，已有研究表明，蛋白的糖基化修饰对于外泌蛋白的功能具有重要意义。本研究利用稻瘟菌野生型菌株 P131 和糖基转移酶突变体 ALG3KO 进行外泌蛋白的定量比较，将二者在 CM 培养基中摇培 18 h 后，提取培养液中的蛋白。通过蛋白质质谱鉴定及无标定量（LFQ）比较分析发现：在 ALG3KO 的突变体中，一部分分泌蛋白的分泌量下降甚至不分泌，另有一部分胞内蛋白被分泌到胞外。本研究初步表明，蛋白的糖基化修饰对于外泌蛋白的正确分泌具有重要作用。后期，我们将筛选具有糖基化修饰位点的外泌蛋白，利用分子生物学手段验证这些蛋白的定位和分泌情况，并鉴定它们的生物学功能。

关键词：稻瘟菌；外泌蛋白；糖基化修饰

通讯作者：卢训莉，副教授，主要从事水稻与稻瘟菌的互作研究；Tel：010-62732541，E-mail：luxunli@cau.edu.cn。

江苏省稻瘟标样中分离真菌的鉴定及基因组进化分析

杜艳，齐中强，俞咪娜，张荣胜，余振仙，刘永锋*

（江苏省农业科学院植物保护研究所，南京 210014）

摘要：稻瘟病(rice blast)是水稻生产上一种毁灭性真菌病害，平均每年对水稻生产造成10%～30%的损失，严重威胁世界粮食生产安全。目前，稻瘟病菌与水稻的互作机制研究较多，而其他梨孢属真菌与水稻间的互作以及其他梨孢属真菌与稻瘟病菌的互作研究较少，这对于建立新的真菌-植物及真菌-真菌互作模式系统以及获得新的抗性资源具有重要意义。由于江苏地处沿海温带亚热带区域，江苏省各稻区杂草分布广泛，且田间杂草种类繁多，常见的种类有禾本科、莎草科和阔叶科杂草等。田间杂草植物易受梨孢属真菌的侵染形成病斑，病斑上的孢子会散落在稻株上，影响稻瘟病的发生。前期实验室分离出3个对丽江新团黑谷不致病的菌株2012-18-1、2012-18-2、2011-122-1，观察发现它们在分生孢子形态和大小、分生孢子梗形态等特性上与稻瘟病菌Guy11较为相近。利用ACTIN、ITS等序列进行比对并构建系统发育树，结果表明不致病菌株18-1、18-2和122为梨孢属真菌，与蓉草上的NI919关系最近。分别将18-1、18-2、122与野生型Guy11菌株接种丽江新团黑谷、日本晴和CO39三个品种水稻穗部，发现18-1、18-2、122均能引起穗颈瘟。进一步杂草寄主测定结果表明，该类不致病菌能引起杂草寄主狗尾草、棒头草、无芒稗和光头稗致病，说明该类群梨孢属真菌寄主广泛且发生普遍。进一步通过基因组测序发现该类真菌18-2基因组大小约41.89 M，进化上与稻瘟病菌70-15亲缘关系较远。将18-2的蛋白序列和参考基因组70-15的蛋白序列进行比对，18-2特有cluster 2 168个，包含2 250个基因，其中253个基因具有分泌蛋白功能，占全部特有基因的11.24%。以上结果表明，18-2等不致病菌株作为一类新种，在稻瘟病的发生过程中起到一定的作用，研究结果可望为解析真菌-植物及真菌-真菌互作模式系统及稻瘟病的防治提供新的思路和方向。

基金项目：国家重点研发计划(2016YFD0300706)；江苏省自主创新资金项目CX15(1054)；国家自然科学基金青年基金31601592

通讯作者：刘永锋，研究员，主要从事植物病害致病机制及其防控技术研究；Tel：025-84391002，E-mail：liuyf@jaas.ac.cn

第一作者：杜艳，副研究员，主要从事稻瘟病菌的致病机理研究；E-mail：dy411246508@126.com。

我国不同产区桃褐腐病菌生物学特性研究

朱薇，谈彬，赵文静，陈丽丽，曹军，纪兆林*，董京萍，徐敬友，童蕴慧

（扬州大学园艺与植物保护学院，扬州 225009）

摘要：桃褐腐病是我国桃产区危害桃果的重要病害，同时在桃果运输、储运期间也易发生，引起果腐，造成严重的经济损失。为有效防控不同桃产区桃褐腐病，本文对来自我国不同桃产区的 15 个代表菌株的生物学特性进行了研究。结果表明，不同产区来源的桃褐腐病菌在菌丝生长特性和分生孢子萌发特性上存在一定的差异：如在生长温度方面，相比于其他菌株来自大连、青岛的菌株在较低温度（15℃）下生长良好，与 20℃ 基本一致，而大连分离株在 30℃ 下生长较差。在光照方面，北京和大连分离菌株在全黑暗和全光照条件下生长情况基本一致，而浙江丽水、四川成都、云南昆明分离菌株在全黑暗条件下生长要好于全光照，山东泰安的 1 分离株在全光照下生长要明显好于全黑暗，河北石家庄分离菌株在全黑暗和全光照下生长状况最差。在 pH 方面，大多数菌株在 pH6 下生长要好于 pH7，但北京 1 分离株 pH7 的生长要明显好于 pH6，而河北石家庄、浙江丽水和北京的 1 菌株在 pH6 和 7 的生长状况基本一致，山东泰安 1 菌株在 pH10 下生长好于其他分离株。虽然存在差异，但各产区分离菌株总体趋势基本一致：病菌在 PDA 培养基、25℃、pH 6.0～7.0 及 12 h 光照 12 h 黑暗交替的条件下生长最好；病菌生长最适碳源是淀粉和乳糖，最适氮源是牛肉浸膏和 KNO3。病菌在 PDA 上易产生分生孢子，分生孢子在 25℃、pH 6.0～7.0 及 12 h 光照 12 h 黑暗交替条件下萌发率最高。总之，不同的营养、温度、光照、pH 等环境条件对桃褐腐病菌菌丝生长和孢子萌发有显著影响，且不同来源菌株存在一定差异。

关键词：桃褐腐病；果生链核盘菌；生物学特性

基金项目：国家现代农业产业技术体系建设专项（CARS-30-3-02）；江苏省农业科技自主创新资金项目[CX（14）2015，CX（15）1020]

通讯作者：纪兆林，副教授，研究方向：植物病害生物防治及分子植病研究；E-mail：zhlji@yzu.edu.cn

第一作者：朱薇，硕士研究生，研究方向：植物病害生物防治；E-mail：1098407067@qq.com。

我国桃褐腐病菌种类及致病性分析

谈彬，朱薇，陈丽丽，赵文静，张权，纪兆林*，董京萍，童蕴慧，徐敬友

（扬州大学园艺与植物保护学院，扬州 225009）

摘要：桃褐腐病在我国乃至全世界范围内广泛发生，造成严重的果实腐烂，导致桃果减产、损失严重。为明确我国桃主产区褐腐病菌的种类，本文从桃主产区采集分离褐腐病菌，并进行了种类鉴定和致病性分析。本文分别从浙江丽水、四川成都，云南昆明、辽宁大连、河北石家庄、北京、山东泰安和青岛等 8 个产区褐腐病果上单孢分离获得 15 个代表菌株。从平板菌落形态上，来自云南的分离株 hfyn 与其他分离株菌落明显不同。15 个菌株 rDNA-ITS 区域 PCR 扩增测序和 BLAST 比对结果显示，来自云南的 1 菌株 hfyn 与新种云南丛梗孢（*Monilia yunanensis*）具有 99%的相似性，而从其他产区分离的 14 个菌株均与果生链核盘菌（*Monilinia fructicola*）高度相似。本文同时采用 Ioos 等、Ma 等、Cote 等设计的特异性引物对上述 15 分离株特定区域（微卫星区、RAPD 差异片段）序列进行 PCR 扩增，结果与上述 ITS 鉴定结果一致。致病性接种实验表明，不同来源的褐腐病菌在桃果上的致病力均有差异，其中来自浙江丽水的菌株 hfzl 和河北石家庄的菌株 hfhs 致病力相对较强，而来自山东泰安的菌株 hfst 的致病力相对较弱。因此，目前我国桃产区褐腐病菌主要以 *M. fructicola* 为主，但不同产区来源的菌株存在致病力的差异。

关键词：桃褐腐病；链核盘菌；ITS；致病性

基金项目：国家现代农业产业技术体系建设专项（CARS-30-3-02）；江苏省农业科技自主创新资金项目[CX(14)2015，CX(15)1020]。

通讯作者：纪兆林，副教授，研究方向：植物病害生物防治及分子植病研究；E-mail：zhlji@yzu.edu.cn

第一作者：谈彬，硕士研究生，研究方向：分子植物病理学研究；E-mail：1076935534@qq.com。

湖北省茄科蔬菜根病发生现状及致病病原种类调查

汪华[1*]，张学江[1]，向礼波[1]，杨立军[1]，王飞[2]，喻大昭[1]

([1]湖北省农业科学院植保土肥研究所/农作物重大病虫草害防控湖北省重点实验/农业部华中作物有害生物综合治理重点实验室，武汉 430064；
[2]湖北省农业科学院经济作物研究所，武汉 430064)

摘要：茄科蔬菜根病是指由土传真菌侵染导致的根腐，基腐和茎腐病害的总称，全国各地均有发生。近年来，受土壤连作，重茬的影响，湖北省茄科蔬菜根病发生率正随着种植时间的延长而增加，以严重影响茄科蔬菜的发展。本文作者分别于 2015 年 8 月至 9 月、2016 年 7 月至 8 月，通过对湖北省利川、长阳、兴山、十堰、神农架、襄阳、东西湖、崇阳等茄科（辣椒/番茄/茄子）蔬菜主产区根病发生面积和严重度进行调查，基本掌握了湖北省主要茄科蔬菜根病的发生现状，调查表明，茄科蔬菜根腐病的发生在不同生态区不同年份的表现存在明显差异。各地样本分离培养结果表明：茄科蔬菜根病是由多种病原侵染所致，主要致病病原有蠕孢菌、丝核菌、镰刀菌、链格孢菌等 4 个属；其中利川番茄、辣椒、东西湖辣椒、兴山辣椒和长阳番茄根病主要病原分离物为镰孢菌，其次为丝核菌，少量蠕孢菌，咸宁崇阳辣椒、神农辣椒根病主要病原分离物为丝核菌，其次为镰孢菌，少量蠕孢菌。

关键词：茄科蔬菜；根病；发生现状；病原菌

基金项目：公益性行业专项"作物根腐病综合治理技术方案（201503112-8）"；国家重点研究项目"种子、种苗与土壤处理技术及配套装备研发"（2017YFD0201600）；湖北省农业科学院青年基金项目"小麦全蚀病农药室内生测方法研究（2013NKYJJ09）"

第一作者：汪华（1978- ），男，湖北广水人，副研究员，主要从事作物根部病害、农药研发及应用研究；E-mail：wanghua4@163.com。

甜瓜种子携带镰刀菌的种类鉴定及致病性研究

马国苹，张小芳，华晖晖，吴学宏*

（中国农业大学植物病理学系，北京 100193）

摘要：甜瓜在热带和温带地区广泛种植。由尖孢镰刀菌甜瓜专化型引起的甜瓜枯萎病是一种重要的真菌病害，于1933年在美国明尼苏达州首次报道，目前该病已在亚洲、欧洲、美洲、非洲等甜瓜种植区普遍发生。近年来，伴随着我国设施栽培技术的不断进步，甜瓜种植面积呈逐年增加的趋势，枯萎病的发生也日趋严重，某些年份发病率高达50%～70%，严重影响产量和品质。茄病镰刀菌可引起甜瓜根腐病，我国于20世纪80年代末首次在喀什地区巴楚县发现甜瓜根腐病病株，90年代初迅速蔓延；近几年，该病在山东和北京等甜瓜产区普遍发生并且有一定的加重趋势，可造成减产20%以上。镰刀菌可以在土壤中存活多年，种子也可以携带多种镰刀菌。1997年Cruz-Chouque等对5000份甜瓜种质资源带菌情况进行检测，发现种子携带镰刀菌的比率为1%。本实验室从32份甜瓜种子上分离得到23株镰刀菌分离物，结合形态学鉴定和分子生物学方法对其种类进行了鉴定，并研究了其致病性以及对甜瓜种子萌发和幼苗生长的影响。结果表明：23个分离物中，4个分离物为尖孢镰刀菌，5个分离物为再育镰刀菌，5个分离物为木贼镰刀菌，9个分离物为轮枝样镰刀菌。上述4种镰刀菌对甜瓜幼苗均有致病性，其孢子悬浮液对甜瓜种子萌发和幼苗生长均有一定的抑制作用。

基金项目：公益性行业（农业）科研专项（201503110-14）

通讯作者：吴学宏，教授，主要从事植物病原真菌种类鉴定及其遗传多样性研究；E-mail：wuxuehong@cau.edu.cn

第一作者：马国苹，博士研究生，研究方向为西甜瓜真菌病害的病原鉴定。

甜瓜种子携带链格孢菌的种类鉴定及其致病性研究

马国苹，韩涛，刘泉，吴学宏*

（中国农业大学植物病理学系，北京 100193）

摘要：甜瓜在中国、美国、西班牙、土耳其和伊朗等国广泛种植。我国主要分布在新疆、甘肃、内蒙古等西北地区，20 世纪 80 年代开始在华北、华中、华南及台湾地区种植。甜瓜作为我国重要的经济作物，周而复始的种植模式导致各种病原菌逐年累积，使病害发生日趋严重，成为严重制约我国甜瓜产业发展的重要因素。甜瓜叶片在整个生长期极易受多种病原真菌侵染引发叶部病害，主要包括链格孢菌引起的甜瓜叶枯病、甜瓜黑斑病和甜瓜大斑病，瓜类尾孢引起的甜瓜斑点病，瓜灰星霉菌引起的甜瓜叶点病以及棒孢霉引起的甜瓜棒孢叶斑病。其中甜瓜叶枯病菌和甜瓜黑斑病菌均以菌丝体或者分生孢子的形式在病残体上越冬，或者分生孢子附着在种子表面越冬；甜瓜大斑病菌以菌丝体和分生孢子分别在病叶组织内外越冬，成为翌年初侵染源，种子是否带菌尚待研究。本实验室自 32 份甜瓜种子上分离得到 14 个链格孢菌分离物，结合形态学鉴定和分子生物学方法对其种类进行了鉴定，并研究了其致病性以及对甜瓜种子萌发和幼苗生长的影响。结果表明：14 个分离物中，8 个分离物为细极链格孢，3 个分离物为交链格孢，3 个分离物为倒果链格孢。除倒果链格孢外，其余两种链格孢菌对甜瓜离体叶片均有致病性，且这三种链格孢菌对甜瓜种子萌发和幼苗生长均有一定的影响。

基金项目：西甜瓜产业技术体系北京市创新团队植保功能研究室（BAIC10-2017）

通讯作者：吴学宏，教授，主要从事植物病原真菌种类鉴定及其遗传多样性研究；E-mail：wuxuehong@cau.edu.cn

第一作者：马国苹，博士研究生，研究方向为西甜瓜真菌病害的病原鉴定。

两种玉米灰斑病菌的致病性差异研究

刘可杰，姜钰，胡兰，徐婧，董怀玉

（辽宁省农业科学院植物保护研究所，沈阳 110161）

摘要：由玉米尾孢（*Cercospora zeina*）引起的玉米灰斑病已被报道在南非、乌干达、美国、赞比亚、津巴布韦和巴西等国家发生，2011—2013 年，笔者在我国的云南、湖北等地区发现该病有不同程度发生，为我国玉米新见病害。玉米种质资源对玉米尾孢的抗性鉴定与评价尚未见报道。2013—2016 年，通过研究试验，建立了玉米尾孢灰斑病的田间人工接种抗病鉴定技术体系，采用此技术评价了 300 份玉米种质对玉米尾孢（*C. zeina*）和玉蜀黍尾孢（*C. zeae-maydis*）的抗性，结果表明：对玉米尾孢灰斑病表现高抗、抗和中抗的种质分别有 10 份、100 份和 121 份，表现感和高感的种质依次有 62 份和 7 份；对玉蜀黍尾孢灰斑病表现高抗、抗和中抗的种质分别有 4 份、33 份和 78 份，表现感和高感的种质依次有 137 份和 48 份。300 份种质中只有 77 份对两种灰斑病的抗性表现一致，而 149 份种质对两种灰斑病的抗性表现截然相反，说明两种灰斑病菌的致病性有很大的差异。76.99%的供试种质对玉米尾孢灰斑病表现为高抗、抗或中抗，而 61.67%的供试种质对玉蜀黍尾孢灰斑病表现为感或高感，说明玉米尾孢对大多数种质的致病力比玉蜀黍尾孢弱。

关键词：玉米灰斑病；抗性评价；玉米尾孢；玉蜀黍尾孢

基金项目：国家科技支撑计划项目（2011BAD16B12、2012BAD04B03、2013BAD07B03）；农业部作物种质资源保护项目（NB2013-2130135-25-15）

第一作者：刘可杰，博士，副研究员，主要从事旱粮作物病害及作物种质资源抗病虫研究；E-mail：liukejie8888@hotmail.com。

实时荧光定量PCR监测黄瓜霜霉菌方法的建立与应用

张艳菊*,付璐,白玲玲,党悦嘉,刘大伟,王春龙,徐天舒,张奥

(东北农业大学农学院,哈尔滨 150030)

摘要:由古巴假霜霉菌〔*Pseudoperonospora cubensis* (Berk. & M. A. Curtis) Rostovzev〕侵染引起的霜霉病是黄瓜上流行性极强的病害,常给黄瓜生产造成毁灭性损失。加强霜霉病的预测和监测对指导田间病害防治具有重要意义。实时荧光定量PCR(Real-time quantitive PCR)是可以对靶标核酸进行定量检测的技术,具有快速、方便、准确等优势,已广泛应用于植物病原菌的检测,但针对黄瓜霜霉病菌的实时荧光定量PCR方法则鲜有报道。本研究根据黄瓜霜霉病菌的ITS-rDNA序列及Beta-tubulin (btub) gene序列设计并筛选出黄瓜霜霉病菌的特异性引物PCi-3(上游引物PCi-3F序列为:5′-TTGCACTTCCGGGTTAGTCC-3′,下游引物PCi-3R序列为:5′-GGTCACATGGACAGCCTTCA-3′);建立了黄瓜霜霉病菌的荧光定量PCR体系为:无菌去离子水7.6 μL,*Taq* SYBR® Green qPCR Premix 10 μL,上下游混合引物(10 μmol/L)0.4 μL,模版2 μL;最佳反应条件为:94℃ 3 min,94℃ 10 min,60℃ 30 s,72℃ 30 s,74℃ 5 min,40个循环。进一步利用该荧光定量PCR体系检测黄瓜霜霉病菌侵染黄瓜子叶不同时段植物组织内病菌潜伏侵染量的动态变化。病菌在10~12 h(Ct值<30时)时开始侵染叶片,接菌后1~9 d,随时间的增长,Ct值呈下降的趋势,菌量及病情级别都呈上升趋势。Ct值<23时,病情级别为1;Ct值<16时,病情级别为2;Ct值<13.5,病情级别为3;Ct值<11,病情级别为4。在黄瓜生产季将传统的孢子捕捉方法与荧光定量PCR相结合监测了田间霜霉病发生与流行,两种方法检测结果基本保持一致,但荧光定量PCR更灵敏、快速。研究结果为黄瓜霜霉病的预测预报和防治提供了技术手段。

关键词:黄瓜霜霉病;实时荧光定量PCR;田间监测

基金项目:黑龙江省科学基金项目(JJ2016ZZ0064)

通讯作者:张艳菊,教授,博士生导师,研究方向为植物病原生物学;E-mail: zhangyanju1968@163.com。

西南地区西瓜蔓枯病菌 ISSR 类群生物学特性分析

谭蕊，黄萌雨，杨宇衡，余洋，毕朝位*

（西南大学植物保护学院，重庆 北碚 400715）

摘要：西瓜蔓枯病是由 *Stagonosporopsis* 属下 3 个形态学不可区分的种引起的世界性病害，在我国西南地区发病严重，对西瓜产量造成严重损失。前期研究表明该地区的西瓜蔓枯病菌主要为 *S. citrulli*，并运用简单序列重复区间(inter-simple sequence repeats，ISSR)分子标记技术，将采自该地区 7 个区县的 186 株 *S. citrulli* 病菌进行了遗传多样性分析，将这些病菌聚类成八个 ISSR 类群。了解这些类群生长、繁殖、致病力之间的差异，清楚其遗传多样性与生物学特性之间的联系，有利于深入了解该病害，以便对病害更好地进行防治与预报。本文测定了这 8 个 ISSR 类群在 PDA 培养基上的生长速率，结果表明各类群生长曲线基本重叠，无显著差异；比较不同温度对各类群的生长影响，结果表明同一温度下，各类群菌丝生长情况表现出相同趋势：都在 25℃ 时生长最快，10℃ 和 30℃ 生长相对较慢，5℃ 和 35℃ 基本不生长；且在相同温度条件下，各类群菌株之间的菌落大小不具有明显差异，说明西瓜蔓枯病菌对温度的耐受力与 ISSR 类群无关。不同 pH 对西瓜蔓枯病菌菌丝生长影响测定结果表明，病原菌在 pH 4～10 都能生长，酸性条件下生长更快，其中 pH 为 7 时，生长最旺盛；相同 pH 条件下各类群菌落大小不具有明显差异，说明 ISSR 类群菌株对培养基酸碱度适应性相同。在磷酸二氢铵琼脂培养基上分生孢子产生能力测定结果表明，类群Ⅱ与类群Ⅲ产孢能力与其他类群达到显著差异水平，类群Ⅱ产孢数最少，仅 $(3.57\pm0.25)\times10^6$ 个/mL，类群Ⅲ产孢数最多为 $(6.47\pm1.74)\times10^6$ 个/mL，其他类群产孢量在 $(4.43\pm0.19)\times10^6$ 个/mL～$(5.77\pm0.32)\times10^6$ 个/mL 之间。分生孢子萌发实验结果表明，8 个类群的孢子在 25℃ 保湿培养 6 h 后几乎全部萌发，且萌发速率曲线大致相同。不同类群菌株致病力测试结果表明，ISSR 类群菌株致病力存在显著差异，其中类群Ⅳ菌株致病力最弱，病斑平均大小仅 (4.60 ± 0.66) cm^2；其他类群之间致病力无显著差异，平均病斑面积在 (5.61 ± 1.25)～(7.10 ± 0.42) cm^2 之间。综上所述，西瓜蔓枯病菌 ISSR 类群菌株菌丝生长速率、对温度和培养基酸碱度的适应性、孢子萌发能力都没有显著性差异；ISSR 类群间的致病力和产孢能力具有一定的差异，但这种差异与菌株分布的地理位置无关。

关键词：西瓜蔓枯病菌；ISSR 类群；生物学特性

基金项目：中央高校基本科研业务费专项资金(XDJK2017B026)；公益性行业(农业)科研专项 (201303023)

通讯作者：毕朝位，男，副教授，主要从事植物真菌病害及病原菌抗药性研究；E-mail：chwbi@swu.edu.cn

第一作者：谭蕊(1992-　)，女，重庆市南岸区人，硕士研究生，主要从事植物真菌病害方面的研究；E-mail：739981030@qq.com。

新疆伊犁河谷锈菌分类的初步研究

王丽丽，黄雅洁，木坎热木·米吉提，切开古里·克孜汗，
禹媛，热吾扎·苏里坦白，李克梅*

（新疆农业大学农学院，农林有害生物监测与安全防控重点实验室，乌鲁木齐 830052）

摘要：伊犁河谷位于新疆天山最西段，三面环山，地处 80°09′-84°56′E、42°14′-44°50′N 之间，辖八县三市。由于湿润的海洋气流从准噶尔盆地西部进入的影响，使伊犁河谷成为全疆最大的降水中心，也是亚欧大陆干旱地带的一块“湿岛”。得天独厚的水土光热资源，使该地区蕴藏的动、植物及微生物生物等资源异常丰富。

2012—2016 年，本课题组在伊犁河谷地区进行锈菌调查采集，初步已对 213 份植物锈菌标本进行形态学分类研究，共鉴定出锈菌 9 属，82 种。其中，新疆锈菌新纪录 4 种：羊角芹柄锈菌 *Puccinia aegopodii*（Strauss）Rohling，；鹅观草柄锈菌 *Puccinia roegneriae* J. Y. Zhuang & S. X. Wei；鲜卑芨芨草柄锈菌 *Puccinia achnatheri-sibirici* Wang，堇菜柄锈菌 *Puccinia viola* de Condolle。涉及寄主植物 25 科，65 属，114 种，其中，国内锈菌寄主植物新纪录 14 种，新疆锈菌寄主植物新纪录 10 种。

初步区系分析表明，伊犁河谷地区植物锈菌以柄锈菌属 *Puccinia* 为主，有 45 种，占总数的 54.9%，为优势种群；其次为单胞锈菌属 *Uromyces* 有 18 种，占 22.0%；栅锈菌属 *Melampsora* 有 6 种，占 7.3%；多胞锈菌属 *Phragmidum* 有 5 种，占 6.1%；胶锈菌属 *Gymnosporangium* 有 3 种，占 3.7%；金锈菌属 *Chrysomyxa* 有 2 种，占 2.4%；长栅锈菌属 *Melampsoridium* 有 1 种，占 1.2%；糙孢锈菌属 *Trachyspora* 有 1 种，占 1.3%；疣双胞锈菌属 *Tranzschelia* 有 1 种，占 1.2%。

从寄主植物分析来看菊科植物 Asteraceae 有 15 属 33 种，占到寄主总数的 28.9%，为伊犁河谷地区植物锈菌主要寄主类群；其次是蔷薇科 Rosaceae 5 属 17 种，占 14.9；豆科 Leguminosae 4 属 9 种，占 7.9%；禾本科 Gramineae 7 属 7 种，占 6.1%；唇形科 Lamiaceae 4 属 5 种，占 4.4%；杨柳科 Salicaceae 2 属 5 种，占 4.4%；牻牛儿苗科 Geraniaceae 1 属 5 种，占 4.4%；伞形科 Umbelliferae 4 属 4 种，占 3.5%；蓼科 Polygonaceae 2 属 4 种，占 3.5%；藜科 Chenopodiaceae 3 属 3 种，占 2.6%；毛茛科 Ranunculaceae 2 属 3 种，占 2.6%；百合科 Liliaceae 2 属 2 种，占 1.8%；凤仙花科 Balsaminaceae 、龙胆科 Gentiana 和鸢尾科 Iridaceae 分别为 1 属 2 种，各占 1.8%；桦木科 Betulaceae、白花丹科 Limonium、柳叶菜科 Onagraceae、茶藨子科 Ribesiaceae、茜草科 Rubiaceae、旋花科 Convolvulaceae、忍冬科 Caprifoliaceae、松科 Pinaceae、小檗科 Berberidaceae、莎草科 Cyperaceae 和堇菜科 Violaceae 植物分别为 1 属 1 种，各占 0.9%。

本研究结果仅是对伊犁河谷地区锈菌的初步报道，随着调查范围的扩大，将会有更多类群和种类被发现。此外，另有个别锈菌尚未鉴定到种，还需进一步借助分子生物学方法鉴定。所有研究标本保藏于在新疆农业大学菌物标本室(HMAAC)。

关键词：新疆；伊犁河谷地区；锈菌；分类鉴定

甜瓜种子携带真菌的初步研究

马国苹,张小芳,华晖晖,吴学宏*

(中国农业大学植物病理学系,北京 100193)

摘要:甜瓜隶属于葫芦科、甜瓜属,别名香瓜、甘瓜、哈密瓜,为一年生匍匐性或攀缘性草本植物。甜瓜果实香甜可口,果肉含有丰富的水分、维生素和碳水化合物,可消暑清热、生津止渴;含有蛋白质、脂肪和无机盐等,能补充人体所需的能量和多种营养元素;含有将不溶性蛋白质转化为可溶性蛋白质的转化酶,有助于营养吸收;瓜蒂中的葫芦素B具有保护肝脏、减轻肝损伤、继而阻止肝细胞脂肪变性以及抑制纤维增生的功能;瓜籽中的蛋白质、纤维和粗油含量分别为25.2%、15.4%和37.8%;瓜籽中有较高的赖氨酸,含硫氨基酸次之,还有大量的矿物质,特别是镁、钾和磷。据联合国粮农组织(FAO)统计,2016年中国甜瓜的种植面积约为43万hm^2,总产量约0.15亿t。近年来,保护地甜瓜连年种植,甜瓜生长环境中病原菌的大量积累,加上保护地内高温、高湿的生长环境,导致甜瓜病害发生日趋严重,严重制约了甜瓜产业的健康发展。国外研究表明,南瓜、黄瓜、西瓜和甜瓜种子共携带有八种真菌,分别是青霉属、链格孢属、镰刀菌属、丝核菌属、根霉属、炭疽菌属、壳球孢菌和黑曲霉。本研究对来自北京、河北、黑龙江、陕西和河南等地的32份甜瓜种质资源进行种传真菌检测。结果表明,干种子、水洗液、水洗后种子和1%NaClO处理后甜瓜种仁均检测到青霉属、曲霉属、镰刀菌属和链格孢属真菌;经1%NaClO处理后的甜瓜种子、种皮和种仁的带菌率明显低于干种子的带菌率,且经1%NaClO处理后的甜瓜种子和种皮未检测到曲霉属真菌。本研究初步明确甜瓜种子的带菌情况及其携带的主要真菌种类,可以为生产中甜瓜品种的选择及进行合适的种子处理提供一定的理论依据和参考。

基金项目:公益性行业(农业)专项(201503110-14);西甜瓜产业技术体系北京市创新团队植保功能研究室(BAIC 10-2017)

通讯作者:吴学宏,教授,主要从事植物病原真菌种类鉴定及其遗传多样性研究;E-mail:wuxuehong@cau.edu.cn

第一作者:马国苹,博士研究生,研究方向为西甜瓜真菌病害的病原鉴定。

农杆菌介导荸荠秆枯病菌遗传转化体系的构建

颜梅新，董伟清，何芳练，江文，欧昆鹏，高美萍，
张尚文，王艳，韦绍龙，蒋慧萍，黄诗宇，黄伟华

（广西壮族自治区农业科学院生物技术研究所）

摘要：荸荠（*Eleocharis dulcis*）俗称马蹄，富含淀粉和矿物质，具有食用和药用价值的草本植物。但荸荠栽培过程中常遭病害侵扰，其中以秆枯病（俗称"荸荠瘟"）发生最为严重，该病主要为害叶鞘、叶状茎、花器等部位，严重影响荸荠的产量和品质。该病病原为荸荠柱盘孢菌 *Cylindrosporium eleocharidis* Lentz，属半知菌类黑盘孢目柱盘孢属。目前国内外关于荸荠秆枯病菌致病基因方面的研究鲜有报道，其主要限制因素之一是该菌的遗传转化方法缺乏。因此，荸荠秆枯病菌的遗传转化体系的建立对于该菌的致病机理研究具有重要意义。本研究参考前人在真菌遗传转化方面的研究成果，利用含有绿色荧光蛋白 GFP 基因和潮霉素抗性基因 hph 的载体通过农杆菌介导对荸荠秆枯病菌进行遗传转化，初步建立了农杆菌介导荸荠秆枯病菌的遗传转化体系。研究结果表明经 PDA 培养基培养 15 d（或燕麦培养基培养 7 d）后收集的荸荠秆枯病菌孢子，配成孢子悬浮液（1×10^6 或 1×10^7 个孢子/mL）与农杆菌菌液各 100 μL 混合后涂板于加入 AS（200 μmol/L）的覆盖有纤维素滤膜的 IM 固体培养基上，24℃下共培养 72 h。转移纤维素滤膜至含有潮霉素 B（100 μg/mL，）和头孢噻肟钠（300 μg/mL）的 PDA 平板，28℃下培养 5～7 d，可成功获得转化子，通过荧光显微镜、PCR 和 southern blotting 等方法验证了转化子。该技术体系的建立为今后荸荠秆枯病菌致病机制研究提供技术支持。

关键词：荸荠；荸荠秆枯病菌；农杆菌介导；遗传转化

基金项目：广西自然科学基金（2016GXNSFAA380006）

第一作者：颜梅新（1978-　），男，博士，副研究员，主要从事特色水生作物病虫害防治研究；E-mail：574252968@qq.com。

真菌自噬基因的保守性研究

王秦虎[1]，江聪[1]，刘慧泉[1]，许金荣[1,2*]

([1]西北农林科技大学植物保护学院，杨凌 712100；
[2]美国普度大学植物及植物病理系，印第安纳州 47907)

摘要：自噬是真核生物维持胞内环境稳定和营养循环利用的基本降解途径。目前为止，已有数百个真菌基因组公开。然而，真菌中是否有一些谱系特异的自噬基因复制和丢失事件尚不清楚。通过分析 41 个自噬基因在 342 个真菌中的保守情况，我们发现，20 个自噬基因在所有真菌中高度保守，7 个仅在子囊菌中保守，而另外 14 个则为酵母纲真菌特有基因。微孢子菌（低等真菌）和肺孢子菌（子囊菌）是典型的活体寄生真菌，其所有自噬基因几乎全部丢失；然而，大多微孢子虫的 *ATG1* 和 *ATG15* 则在进化中保留了下来。一些可降解木材的木腐菌（主要为担子菌），其自噬基因也大量丢失。马拉色菌（担子菌）可引起毛囊炎、溢脂性皮炎和头皮屑等，其自噬基因数目大量减少。毛霉菌亚门真菌是早期分化的真菌之一，其 7 个自噬基因曾发生了全基因组的复制。绝大多数真菌只编码一个自噬标记基因 *ATG8*；然而，在引起脚气和灰指甲等人类皮肤病的皮肤癣菌中，*ATG8* 基因发生了复制且两个 Atg8 存在功能分化。对自噬重要复合体的保守性分析表明其均存在一个保守组分和一个非保守组分。此外，我们还发现大约有 90% 的禾谷镰刀菌和脉胞菌的自噬基因在其有性生殖阶段发生了 RNA 编辑。

关键词：真菌；自噬基因；基因家族；基因复制；基因丢失；RNA 编辑

通讯作者：许金荣，教授；E-mail：jinrong@purdue.edu。

小麦赤霉病菌病毒多样性研究

花翔旻[1,2],谢甲涛[1],程家森[1],陈桃[1],付艳苹[1],姜道宏[1,2*]

([1]湖北省作物病害监测和安全控制重点实验室(华中农业大学),武汉 430070;
[2]农业微生物学国家重点实验室(华中农业大学),武汉 430070)

摘要:小麦赤霉病(Fusarium head blight)是由禾谷镰刀菌(*Fusarium graminearum*)引起的,危害小麦生产最为严重的病害之一。在我国长江中下游冬麦区、东北春麦区以及华北南麦区发病尤为严重,地理分布广泛,现阶段对于小麦赤霉病的防治手段主要为化学防治,容易导致农药残留、环境污染、引起病菌抗药性等一系列问题。真菌病毒可用作对植物病害进行生物防治的重要手段之一,如本实验室发现的真菌首例 DNA 病毒 SsHADV-1 可在油菜田间降低菌核病发病率。之前研究报道禾谷镰刀菌自然群体中含有种类丰富的病毒,且发现具有弱毒特性的 Fusarium graminearum hypovirus 1(FgHV1)和 FgHV2,这些病毒可抑制镰刀菌的有性生殖,减少产孢,减弱菌丝生长速率和其致病力。本研究通过采集、分离和纯化来自湖北省随州市、广水市、孝感市以及江苏省南京市的小麦麦穗,得到了 400 多株小麦赤霉病菌样本,对其进行转录组测序,分析得到 25 种不同的病毒信息。从中挑选出未曾在小麦赤霉病菌中报道过的两个分别属于 Phlebovirus 和 Sclerotimonavirus 病毒属负链病毒。用特异性引物进行定位,确定小麦赤霉病菌菌株 XG-7 和 SZ-111 携带有以上两种病毒,将两个病毒分别命名为 FgPV 和 FgNSRV。FgPV 有三个 RNA 片段,基因组总长 11~19 kb,测序已获得编码 RdRp 的 L 片段部分信息,而 S 和 M 片段信息未能在测序中得到。FgNSRV 仅有一个 RNA 片段,编码 RdRp 和三个功能未知的蛋白。关于两个病毒基因组的全长克隆以及含毒菌株相关生物学特性的研究仍在继续。

通讯作者:姜道宏,教授,主要从事分子植物病理学及生物防治的相关研究;E-mail:daohongjiang@mail. hzau. edu. cn
第一作者:花翔旻,男,在读硕士研究生,主要从事分子植物病理学相关研究;E-mail:bloom@webmail. hzau. edu. cn。

我国不同地区黄瓜花叶病毒的变异分析

于成明，刘珊珊，亓哲，耿国伟，原雪峰*

（山东农业大学植物保护学院植物病理系，
山东省农业微生物重点实验室，泰安 271018）

摘要：黄瓜花叶病毒（*Cucumber mosaic virus*，CMV）是雀麦花叶病毒科（*Bromoviridae*）黄瓜花叶病毒属（*Cucumovirus*）成员，基因组由 RNA1、RNA2、RNA3 三条正义单链 RNA 组成。黄瓜花叶病毒寄主广泛，传播途径多样，是一种世界性分布的病毒。本研究通过对来自全国不同地区和不同寄主 CMV 分离物 RNA3 的部分序列进行克隆及对比，对我国不同地区黄瓜花叶病毒进行差异性研究。主要通过 RT-PCR 技术和 Western Blot 技术分别对哈尔滨、成都、泰安、烟台四个地区的包括果树、花卉、杂草等在内的 20 多种不同寄主待检样品进行 CMV 检测。其中 RNA3 的检测片段是 119-768，Western blot 使用 CMV 的 CP 蛋白多克隆抗体进行检测。检测结果显示：哈尔滨梨树、四季海棠，成都牛矢果、樱花，烟台大樱桃，泰安杏树、海棠树以及多种杂草中都检测到了 CMV 的存在。序列分析对比发现，不同地区的 CMV 分离物存在较大的区域差异，成都地区牛矢果 CMV 分离物与泰安地区杏树以及哈尔滨地区梨树的 CMV 分离物的相似度只有 81.25% 和 82.16%；而同一地区不同寄主的 CMV 序列差异不大，泰安地区板栗和山枣分离物相似度为 99.8%。

关键词：黄瓜花叶病毒；序列分析；变异

基金项目：国家自然科学基金资助项目（31670147；31370179）；山东省“双一流”建设资助经费（SYL2017XTTD11）

通讯作者：原雪峰，教授，博士生导师，主要从事分子植物病毒学研究；E-mail：snowpeak77@163.com

第一作者：于成明，山东平阴人，博士后，主要从事分子植物病毒学研究；E-mail：ycm2006.apple@163.com。

来源于湖北地区的梨轮纹病菌菌株携带真菌病毒多样性的研究

罗慧[1,2],王利华[1,2],胡旺成[1,2],洪霓[1,2],王国平[1,2],王利平[1,2]*

([1]华中农业大学植物科学技术学院,
湖北省作物病害监测与安全控制重点实验室;
[2]华中农业大学,农业微生物学国家重点实验室,武汉 430070)

摘要:梨轮纹病是主要由葡萄座腔菌(*Botryosphaeria dothidea*)引起的一类真菌病害,在我国梨产区以及栽培园中均有感染,造成梨树枝干、果实产生轮纹状病斑,通常与腐烂病和干腐病混合感染导致梨树衰退和死亡。真菌病毒(mycovirus)是一类以真菌为寄主的病毒,它在真菌中普遍存在,大多数对病原真菌的营养生长,菌落形态,致病力等无影响,也有少部分对真菌的生长速率,产孢能力,致病力等生物学特性有严重影响。目前,应用低毒相关病毒防治病原真菌是研究的热点,如板栗疫病和油菜菌核病,均对真菌病害起到很好的生防效果。随着梨轮纹病危害逐年加重,寻找新的、安全有效的措施防治梨轮纹病显得尤为迫切,弱毒相关病毒导致病原真菌致病力衰退,可作为果树病害生防的另一条重要途径。

本实验室随机采集来源于"湖北省武汉市果树茶叶研究所砂梨种质资源圃"发病梨树枝干病样,对其进行分离和纯化,经鉴定共获得了13份梨轮纹病菌菌株,对其进行离体接种"皇冠"梨果和"丰水"叶片,测定其致病力。结果表明,13份分离株均能在梨果和叶片上形成典型的黄褐色或深褐色大小不同的轮纹状病斑,表明来源于湖北地区的梨轮纹病菌菌株对其寄主梨具有不同程度的致病性。对13份分离株进行dsRNA提取和检测,发现12份菌株均携带分子量不同大小的dsRNA片段,感染真菌病毒率为92.3%(12/13)。对13份分离株携带dsRNA片段大小及带型进行观察,3份分离株携带的dsRNA带型一致,均含有分子量在1~7 kb之间的7条dsRNA条带,初步推测复合感染2种以上真菌病毒。其他9份菌株的dsRNA条带位置均位于以上分离株携带7条带型中的某一条或2条条带以上带型位置。以上结果初步表明,来源于湖北地区梨轮纹病菌携带真菌病毒具有多样性,将其dsRNA进行cDNA克隆和测序,明确其病毒种类,为真菌病毒的分类提供新资源,也为果树轮纹病害生物防治提供真菌病毒资源。

关键词:梨轮纹菌;真菌病毒;多样性

基金项目:国家自然科学基金(31471862);国家梨产业技术体系(CARS-29-10)

通讯作者:王利平,副教授,研究方向为果树病理学;Tel:+86-18696119086,Fax:+86-27-87384670,E-mail:wlp09@mail.hzau.edu.cn。

枣树花叶相关病毒的全序列扩增和分析

杜开通[1]，刘思佳[1]，田国忠[2]，周涛[1*]

([1]中国农业大学植物病理学系，农业部植物病理学重点实验室，北京 100193；
[2]中国林业科学研究院，国家林业局森林保护学重点实验室，北京 100091)

摘要：枣树在我国栽培历史悠久，分布范围广，是我国最具经济价值的特色树种之一。枣树花叶病在我国大部分枣区普遍发生，是一种长期困扰枣树生产的重要病害。以采集自北京市海淀区中国林业科学院科研基地的典型花叶病症状的枣树叶片为材料，提取总 RNA 进行深度测序，深度测序数据经生物信息学分析，发现样品中很可能存在杆状 DNA 病毒属病毒(badnaviruses，family *Caulimoviridae*)。利用杆状 DNA 病毒属简并引物 BadnaFP/BadnaRP 对花叶样品总 DNA 进行 PCR 验证，扩增得到约 580 bp 的特异条带，克隆测序后经 BLASTN 比对分析表明其可能为一种新的 badnavirus，将其命名为枣树花叶相关病毒(Jujube mosaic-associated virus，JuMaV)。在已获得的 580 bp 序列上设计反向长距离引物 IF/IR，扩增到约 7200 bp 大小的条带，连接到 pMD-19T，克隆后测序。再在 IF/IR 两侧设计引物 F2/R2 以覆盖反向引物所在区域，验证了反向引物扩增到的是环状 DNA，目的条带经测序，最终拼接得到 7194 bp JuMaV 基因组全序列。JuMaV 基因组正链包括 5 个开放读框(open reading frames，ORFs)，其中 ORF1、ORF2、ORF4 与已报道 badnaviruses 的 ORFs 在基因组上的位置及大小相似，但 JuMaV 的 ORF3 具有 70 nt 的基因间隔区，将移动蛋白和复制功能区隔开，分别命名为 ORF3a 和 ORF3b。经逆转录(RT)和 RNA 酶 H(RNase H)区氨基酸序列进化树分析，发现 JuMaV 与 badnaviruses 归为一组，但 RT/RNase H 区与 badnaviruses 相似性低于 80%，且基因组序列相似性低于 50%。因此，JuMaV 是 *Caulimoviridae* 的一个新成员。

基金项目：111 专项 B13006

通讯作者：周涛，副教授，主要从事植物病毒学研究；E-mail：taozhou@cau.edu.cn。

一种菊花新病毒的鉴定

王蓉[1]，董佳莉[1]，岳瑾[2]，周涛[3]，李勇[1]，丁万隆[1*]

([1]中国医学科学院北京协和医学院药用植物研究所，北京 100193；
[2]北京市植物保护站粮经作物科，北京 100029；
[3]中国农业大学植物病理学系，农业部植物病理学重点实验室，北京 100193)

摘要：菊花(*Dendranthema morifolium*)是菊科(Asteraceae)菊属(*Dendranthema*)的多年生草本植物，不仅具有观赏价值，还是我国传统的中药材之一，具有散风清热、解毒消炎等功效。菊花的繁殖以扦插为主，易造成病毒病的快速传播。目前，已报道侵染菊花的病毒有 20 余种。2017 年初，在北京市延庆区发现表现矮化和轻微花叶症状的菊花植株，疑似病毒侵染。利用 RT-PCR 方法，未检测到可引起菊花矮化或花叶症状的病毒。为鉴定表现病毒病症状的菊花是否携带病毒以及病毒的种类，本研究通过 Illumina 高通量测序的方法对菊花小 RNA 文库进行测序，对测序结果分析发现：其中 26 条长度 46～183 bp 的 contigs 与香石竹潜隐病毒属(*Carlavirus*)成员的核苷酸相似度较高(74.04%～88.89%)，初步推测为一种新病毒。根据 contigs 序列设计引物，经 RT-PCR 和 RACE PCR 扩增菊花新病毒的全基因序列。该病毒基因组全长 8 874 bp(不含 poly(A))，其基因组结构与香石竹潜隐病毒属成员相似，编码 6 个蛋白，5′端有磷酸帽子结构，3′端有 poly(A)尾。新病毒的多聚蛋白序列与菊花 B 病毒(*Chrysanthemum virus B*)Uttarakhand 分离物(GenBank accession No. AM765838)的相似度最高，核苷酸水平为 65.57%，氨基酸水平为 67.07%，外壳蛋白序列与水仙普通潜隐病毒(*Narcissus common latent virus*)Zhangzhou 分离物(AM158439)的相似度最高，核苷酸水平为 55.99%，氨基酸水平为 52.27%，达到 ICTV 分类标准中对于认定新种的标准，暂命名为菊花 R 病毒(Chrysanthemum virus R)。

基金项目：中国医学科学院医学与健康创新工程(2016-I2M-3-017)

通讯作者：丁万隆，研究员，主要从事药用植物栽培研究；Tel：010-57833360；E-mail：wlding@implad.ac.cn
第一作者：王蓉，助理研究员，主要从事药用植物病毒病相关研究；E-mail：wangrong1019@163.com。

山东樱桃病毒病病原分析

曹欣然[1,2]，刘珊珊[1]，耿国伟[1]，亓哲[1]，于成明[1]，原雪峰[1*]

([1]山东农业大学植保学院 山东省农业微生物重点实验室，泰安 271000；
[2]烟台市农业技术推广中心，烟台 264001)

摘要：山东省是我国樱桃栽培面积最大的省份，栽培总面积约 6 万 hm^2，主要集中在烟台福山、泰安岱岳等地。樱桃在其长期无性繁殖和栽培过程中，大多感染并积累了多种病毒，病毒病已成为制约樱桃产业发展的重要因素之一。本研究在山东省樱桃主产区烟台和泰安广泛采集样品，共获得 44 份混合样品（每个县市区选择樱桃主栽区，每个主栽区选择 5 个代表性果园，一个县市区为一个混合样品）。纪录采集样品的品种、表现症状、地点及树龄。利用 RT-PCR 技术进行病毒病种类检测。采用去多糖多酚 TransZol 提取植物总 RNA，根据文献报道针对 17 种病毒设计特异性引物进行检测。结果表明：山东樱桃主产区烟台、泰安果园均有病毒病发生，感病品种多为红灯、美早，老果园病毒病发生严重。主要表现症状为叶片皱缩、斑驳、畸形、边缘缺刻、花叶、蜷曲、狭长，枝条短缩，果实斑驳（花果），开花后不坐果，坐果少或果小等症状。共检测出 8 种病毒，分别是 CVA（*Cherry virus A*）、CGRMV（*Cherry green ring mottle virus*）、LchV-1（*Little cherry virus 1*）、LchV-2（*Little cherry virus 2*）、PBNSPaV（*Plum bark necrosis stem pitting-associated virus*）、PNRSV（*Prunus necrotic ringspot virus*）、PDV（*Prune dwarf virus*）、CMV（*Cucumber mosaic virus*）。采集样品中，复合侵染率为 100%；其中 CVA 检出率高达 100%，其次是 CGRMA 检出率为 58.8%，CMV 检出率为 35.3%，PBNSPaV 检出率为 32.4%，PNRSV 和 PDV 检出率相同为 23.5%，LchV-1 检出率为 5.9%，LchV-2 检出率最低为 2.9%。采集样品中，一个樱桃样品中最高同时感染 7 种病毒，最少同时感染 2 种病毒。本研究为山东地区樱桃病毒病的发生情况和病原鉴定提供了依据，为田间防治病毒病提供了理论依据。

关键词：樱桃；病毒病；复合侵染

基金项目：国家自然科学基金资助项目（31670147；31370179）；山东省“双一流”建设资助经费（SYL2017XTTD11）
通讯作者：原雪峰，教授，博士生导师，主要从事分子植物病毒学研究；E-mail：snowpeak77@163.com
第一作者：曹欣然，女，在读博士研究生，植物病毒学研究；E-mail：xinran1001@163.com。

贵州省太子参主要病毒病的分子检测及鉴定

王廿[1]，董佳莉[2]，黄刚[3]，陶刚[1*]，王蓉[2*]

([1]贵州省农业科学院植物保护研究所，贵阳 550006；
[2]中国医学科学院药用植物研究所，北京 100103；
[3]黔东南民族职业技术学院，凯里 556000)

摘要：太子参(*Psemdostellaria hererophylla*)系石竹科异叶假繁缕属植物，又名孩儿参，是我国传统滋补强壮的中药材，以块根入药。太子参病毒病是危害太子参生产的重要病害之一，长期以来以农户多用自留种根，通过无性繁殖获得种苗，如所用种根受到病毒侵染，导致种苗带毒，增加了田间病毒病的发生，据统计，由病毒病引起的太子参产量损失在10%～30%。目前在太子参病毒研究中报道了4种病毒，包括烟草花叶病毒(*Tobacco mosaic virus*, TMV)、芜菁花叶病毒(*Turnip mosaic virus*, TuMV)、黄瓜花叶病毒(*Cucumber mosaic virus*, CMV)和蚕豆萎蔫病毒2号(*Broad bean wilt virus 2*, BBWV2)。贵州省是我国太子参的主产区之一，为了明确贵州省太子参的病毒种类，在施秉县和黄平县等贵州省重要产区进行田间太子参病毒病调研。采集叶片出现皱缩、褪绿或花叶等类似病毒病症状的植株共计30株，将其中2株症状严重的叶片提取RNA后，通过小RNA测序分析所带病毒种类，结合RT-PCR检测上述4种病毒，检出率达83.3%，共检出TuMV、TMV和BBWV2三种病毒，带毒率依次为70.0%、23.3%和20.0%；出现TuMV和TMV及TuMV和BBWV2的复合侵染，带毒率分别为23.3%和10.0%，三种病毒复合侵染率仅为0.07%，未在检测样品中发现CMV侵染。本研究为贵州省太子参脱毒种苗病毒检测和太子参病毒病的田间防控提供了重要参考信息。

关键词：太子参；分子检测；烟草花叶病毒；芜菁花叶病毒；蚕豆萎蔫病毒2号

通讯作者：陶刚，博士，主要从事为植物真菌病害生物防治研究；E-mail：ttg729@sina.com
王蓉，博士，主要从事植物病毒学研究；E-mail：wangrong1019@163.com
第一作者：王廿，博士，主要从事植物病毒学及植物病害生物防治研究；E-mail：jessiewang323@163.com。

番茄“酱油果”病害病毒病原的初步鉴定

刘思佳，杜开通，范在丰，周涛*

（中国农业大学植物病理学系，农业部植物病理学重点实验室，北京 100193）

番茄是我国主要的蔬菜和经济作物之一，病毒病一直是影响番茄高产、稳产的主要限制因素。目前番茄上发生的主要病毒有烟草花叶病毒（*Tobacco mosaic virus*，TMV），番茄花叶病毒（*Tomato mosaic virus*）、烟草轻绿花叶病毒（*Tobacco mild green mosaic virus*）、黄瓜花叶病毒（*Cucumber mosaic virus*，CMV）、番茄黄化曲叶病毒（*Tomato yellow leaf curl virus*），番茄褪绿病毒（*Tomato chlorosis virus*），南方番茄病毒（*Southern tomato virus*，STV）、番茄斑萎病毒（*Tomato spotted wilt virus*）等。引起的症状主要有花叶、蕨叶、卷叶、黄化曲叶、植株矮化以及果实坏死等。番茄感染病毒病常年减产 20%至 30%，流行年份高达 50%至 70%，一些地区甚至绝产，经济损失严重，并影响食品安全。快速、准确地鉴定发病植物上的病毒种类是病毒病害防控的前提。2016 年在北京市郊区县进行番茄病毒病害调查时，发现昌平、顺义等地的温室番茄上发生果实表面大面积坏死、环状坏死斑等症状，当地技术员和菜农俗称之为“酱油果”，疑似病毒侵染引起。本文采用深度测序技术结合 RT-PCR 方法初步鉴定了“酱油果”番茄中的病毒，为此类病害的防控提供参考。

提取番茄果肉总 RNA，利用小 RNA 深度测序技术初步鉴定“酱油果”番茄中的病毒。深度测序后拼接得到的 Contigs 经 local BLASTn 分析，发现存在 STV、TMGMV 和 CMV 复合侵染。为了验证深度测序的结果，设计引物，利用 RT-PCR 方法进行检测。利用 STV-CP 引物，扩增到大小约为 550 bp 的特异性条带；利用 TMGMV-CP 引物，扩增到大小约为 500 bp 的特异性条带；利用 CMV-CP 引物，扩增到大小约为 620 bp 的特异性条带。将 PCR 产物测序，经 BLAST 分析：测序得到的 STV CP 核苷酸序列与 STV 西班牙分离物、美国分离物、墨西哥分离物 CP 序列相似性为 100%；测序到的 TMGMV CP 核苷酸序列与厦门分离物的相似性为 99.2%；测序得到的 CMV CP 核苷酸序列与 ZmBJ-CMV CP 相似性最高为 98%。以上结果初步确定了“酱油果”番茄被 STV、TMGMV 和 CMV 复合侵染。将通过生物学接种以及构建三种病毒侵染性克隆，在番茄上接种三种病毒，观察发病症状，进行柯赫氏法则验证，最终确定引起番茄“酱油果”症状的病毒。

基金项目：北京市现代农业产业技术体系果类蔬菜创新团队

通讯作者：周涛，主要从事植物病毒病害相关研究；E-mail：taozhoucau@cau. edu. cn。

丁香假单胞杆菌 *Pst* DC3000 Ⅲ型分泌系统效应蛋白组学与植物互作研究

魏海雷*

（农业部农业微生物资源收集与保藏重点实验室，
中国农业科学院农业资源与农业区划研究所，北京 100081）

摘要：Ⅲ型分泌系统（Type Ⅲ Secretion System，T3SS）是许多动植物病原细菌主要的致病系统，其结构组成和基因功能具有很高的保守性。对 T3SS 的研究有助于解析病原菌的致病过程，为推动动植物病害的治疗和防治具有重要的科学价值。T3SS 由一个 20～30 kb 的基因簇组成，编码类似于注射器的针状结构，负责将致病和免疫相关的效应蛋白（Effector，T3E）分泌到细菌胞外并向宿主细胞转运。T3E 被认为是病原菌与宿主互作的主要因子。病原细菌具有多个甚至几十个 T3E，共同参与致病和免疫过程。但是，单个 T3E 编码基因的突变对致病的影响微乎其微，因此 T3E 之间的冗余（Effector redundancy）现象成为制约致病性研究的最大瓶颈。为了突破这一障碍，本研究以典型 T3SS 植物病原细菌丁香假单胞菌（*Pseudomonas syringae* pv. *tomato* DC3000，*Pst* DC3000）为模式：①在保留完整的 T3SS 结构基因和特性的基础上，构建了全部 36 个 T3E 编码基因的多聚突变体 D36E；②利用 Gateway 与 Tn7 单拷贝稳定融合技术构建了 T3E 高通量克隆与表达载体。D36E 与 Gateway-Tn7 相结合的体系可以实现：①高通量、快速、便捷的基因克隆；②T3SS 天然诱导启动子转录；③T3E-HA/Cya 融合表达与检测；④单、高效的质粒结合转移；⑤基因组单拷贝稳定整合，从而使目标 T3E 最大程度的接近天然表达状态。这一体系具有 5 个独特的优势：①具备 T3E 表达调控的所有元件，是一个稳定的天然表达宿主；②可以通过分泌与转运准确鉴定 T3E；③可以检测和分析 T3E 激发或抑制免疫活性的能力；④可以解析 T3E 之间的互作关系；⑤可以检测和鉴定 T3E 的致病组合，获得核心效应蛋白组。利用这一体系，我们在 D36E 系统中逐个表达了 29 个高活性 T3E，结果显示：①有 5 个效应蛋白可以在本生烟（*Nicotiana benthamiana*）上完全抑制鞭毛蛋白引起的 PTI（Pattern-triggered immunity）基础抗性，有 6 个效应蛋白完全没有抑制 PTI 的活性；②有 6 个效应蛋白可以引起本生烟的 ETI（Effector-triggered immunity）细胞快速死亡（Cell death），其中 2 个被鉴定为无毒基因编码蛋白（HopQ1-1，HopAD1）；③通过高通量交叉筛选发现，效应蛋白 HopI1 特异性的抑制 HopQ1-1 引起的细胞死亡，而 AvrPtoB 特异性的抑制 HopAD1 引起的细胞死亡；④对效应蛋白进行随机整合，发现 8 个效应蛋白（AvrE1，HopM1，HopAA1-1，HopN1，HopE1，HopG1，HopAM1-1，AvrP-

通讯作者：魏海雷，男，研究员；E-mail：weihailei@caas.cn。

toB)的组合就可以使 D36E 恢复致病力，这是目前已发现的最小致病力效应蛋白组合。综上所述，D36E 与 Gateway-Tn7 结合的体系是目前为止研究效应蛋白最高效、最天然的体系，开创性地实现了效应蛋白组学的系统研究。利用这一新的体系，不但可以实现对近缘菌效应蛋白组的综合测定和评价，还可以更深入的了解 T3E 如何系统性参与致病和免疫两大生物过程。同时，本研究的策略与方法还将为其他细菌性病害提供研究思路和应用平台，为进一步系统解析植物与病原细菌互作提供帮助。

广东茄科青枯菌种群遗传多样性分析

佘小漫[1,2]，蓝国兵[1]，汤亚飞[1]，于琳[1]，何自福[1*]

([1]广东省农业科学院植物保护研究所，广州 510640；
[2]广东省植物保护新技术重点实验室，广州 510640)

摘要：由茄科青枯菌（*Ralstonia solanacearum*（Smith）Yabuuchi *et al*.）侵染引起的植物青枯病是一种难以防治的维管束病害，该病在广东省发生严重，每年均造成严重的经济损失。1999-2016 年期间，本团队共采集各类植物青枯病样 930 份，分离获得青枯菌 769 株。本研究对其中分离自 16 种作物的 202 个代表菌株的生化变种、演化型及序列变种进行鉴定。生化变种测定结果显示，1 株为生化变种 1，5 株为生化变种 2，94 株为生物变种 3，103 株为生化变种 4，生化变种 3 和 4 仍为广东省青枯菌优势生化变种。复合 PCR 鉴定结果表明，197 株可以同时扩增获得 144 bp 和 280 bp 的特异条带，属于演化型 I；而分离自马铃薯的 5 个菌株可以同时扩增获得 280 bp 和 372 bp 的特异条带，属于演化型 II。内切葡聚糖酶基因（*egl*）部分序列的分析结果表明，202 个菌株分属 10 个序列变种，分别为序列变种 1、13、14、15、17、18、34、44、45 和 48，其中序列变种 15、34 和 44 是优势序列变种。分离自番茄的菌株存在最多的序列变种类型，包括序列变种 13、14、15、17、18、34、44 和 48；其次分别为分离自马铃薯、砂姜、茄子和辣椒菌株，各自存在 6 个、5 个、4 个和 3 个序列变种。这些研究结果表明，广东茄科青枯菌种群存在丰富的遗传多样性，为进一步研发广东作物青枯病防控技术提供了科学依据。

关键词：青枯菌；生化变种；演化型；序列变种

基金项目：广东省科技计划项目（2015B020203002，2015A020209057）；广东省农业科学院院长基金（201513）
通讯作者：何自福（1966-　），男，研究员，博士，主要从事蔬菜病理学研究；E-mail：hezf@gdppri.com
第一作者：佘小漫（1981-　），女，副研究员，硕士，主要从事植物病原细菌研究。

河北省甘薯细菌性软腐病病原鉴定

高波,王容燕,马娟,李秀花,陈书龙*

(河北省农林科学院植物保护研究所,
河北省农业有害生物综合防治工程技术研究中心,
农业部华北北部作物有害生物综合治理重点实验室,保定 071000)

摘要:甘薯(*Ipomoea batatas* Lam.)在我国是继小麦、水稻、玉米、大豆之后的第五大粮食作物。甘薯细菌性病害是近几年危害我国甘薯生产的一类重要的病害,严重威胁着甘薯产业的健康发展。目前国内外报道的甘薯细菌性病害主要包括甘薯瘟病和甘薯茎腐病,其中甘薯茎腐病发生最为广泛,已在我国的南北方薯区均有报道。近几年河北省部分甘薯种植区甘薯呈细菌性腐烂的情况时有发生,我们于2015—2016年在对河北省甘薯病害调查中,在大田多次发现甘薯软腐状病样,其症状在'龙薯9'上主要表现为薯块呈水浸状腐烂,变软,伴有恶臭味,与之前报道的甘薯茎腐病的症状相似。经对元氏和雄县采集的样品进行室内病原分离纯化,最终获得了三株代表性菌株DD1、YLT2和XX3,三株菌均能在CVP(crystal violet pectate)培养基上产生圆形凹陷。将三株菌分别回接到薯块上,以接种无菌水为对照,28℃灭菌沙土中保湿培养,24 h后观察发现所有实验组均有腐烂病斑产生且与田间症状一致,而对照组无症状,说明三株菌均为病原菌。通过Biolog微生物鉴定系统(ver 5.2.01, Hayward, CA)分析发现DD1、YLT2和XX3与*Pectobacterium carotovorum* subsp. *carotovorum* (*Pcc*)最为相似,SIM值分别为0.638、0.610和0.840。进一步通过对DD1、YLT2和XX3的16S rDNA(对应GenBank登录号为KU180445、KU180448和KU180447)、*acnA*(KU882668、KU882672和KU882671)和*mdh*(KU882675、KU882674和KU882676)基因的克隆和序列分析结果显示,DD1和XX3与*Pcc*的菌株同源性最高为99%~100%,而YLT2则与*P. carotovorum* subsp. *odoriferum*(*Pco*)的菌株同源性最高为99%~100%,该结果与微生物鉴定系统的结果有所差异。为了进一步验证分子鉴定的结果,通过利用一种专门用于*Pectobacterium*亚种鉴定的生理生化检测方法(Schaad et al. 2001)最终确定DD1和XX3为*Pcc*,而YLT2为*Pco*,即与分子鉴定的结果一致。由此确认导致河北元氏和雄县甘薯软腐状的病原为*Pcc*和*Pco*,此类病原在甘薯上的危害在我国尚属首次报道,该病害暂命名为甘薯细菌性软腐病。

关键词:甘薯;细菌性软腐病;*Pectobacterium carotovorum* subsp. *carotovorum*;*P. carotovorum* subsp. *odoriferum*

基金项目:国家甘薯产业技术体系(CARS-10-B16);河北省自然科学基金(C2015301043);国家自然科学基金(31501616;31401721)

通讯作者:陈书龙,研究员,主要从事植物寄生线虫学研究;E-mail: chenshulong@gmail.com

第一作者:高波(1985-),男,助理研究员,主要从事甘薯病害研究;E-mail: gaobo89@163.com。

高效离子交换色谱法分析青枯雷尔氏菌 Tn5 转座子突变菌株的异质性

郑雪芳，刘波*，朱育菁，陈德局，肖荣凤

（福建省农业科学院农业生物资源研究所，福州 350003）

摘要：由青枯雷尔氏菌（*Ralstonia solanacearum*）引起作物青枯病，是一种毁灭性土传病害，难以防治。利用无致力青枯雷尔氏菌防治作物青枯病具有很好的应用潜力。作者前期利用 Tn5 插入诱变获得一批不同致病力的青枯雷尔氏菌突变菌株。为了研究青枯雷尔氏菌 Tn5 转座子突变菌株的异质性，筛选出无致病力高效生防菌株。本研究采用高效离子交换色谱和紫外检测系统建立快速分离青枯雷尔氏菌的细菌色谱方法；并利用该方法对不同致病力 Tn5 转座子突变菌株的色谱行为进行研究；构建色谱效价指数（Chromatography titer index，CTI_i），$CTI_i = S_i/(S_1 + S_2 + S_3) \times 100\%$（$i = 1$、2 或 3；$S_1$、$S_2$ 和 S_3 分别对应 P_1、P_2 和 P_3 的峰面积），测定不同致病力青枯雷尔氏菌突变菌株的 CTI_i 值，用皮尔逊相关系数（Pearson correlation coefficient，PCC）分析色谱效价指数与突变菌株的致病力和青枯病害防效的相互关系。结果表明，青枯雷尔氏菌的最佳色谱分离条件为：缓冲液 pH 为 8.0，流速为 2 mL/min，菌体浓度要大于 10^8 数量级且小于 10^{11} 数量级。不同致病力青枯雷尔氏菌的高效离子交换色谱分离结果显示，青枯雷尔氏菌经高效离子交换色谱可以分离出 P_1（出峰时间为 0.6 min）、P_2（出峰时间为 4.4 min 或 4.5 min）和 P_3（出峰时间为 5.9 min 或 6.0 min）；强致病力菌株和无致病力菌株均具有 3 种不同峰类型：单峰、双峰和三个峰，强致病力菌株的主峰为 P_3 峰，无致病力菌株的主峰为 P_1 峰，中间型菌株有双峰和 3 个色谱峰两种类型；强致病力菌株中 CTI_3 为 100%的菌株，致病力最强，诱导番茄植株发病率均达 85%以上，$CTI_3 < 100\%$的菌株，接种后番茄植株发病率介于 66.33%～84.14%；无致病力菌株中 CTI_1 达 100%的菌株，其防效高，均达到 80%以上，CTI_1 值低于 90%的菌株，其防效低，介于 16.68%～63.62%；CTI_3 与突变菌株致病力呈极显著正相关，相关系数 PCC 值为 0.62；CTI_1 与无致病力突变菌株的防效呈极显著正相关，相关系数 PCC 值为 0.80。

苦瓜枯萎病生防芽孢杆菌的分离鉴定

郭康迪，赵莹，杨梦珂，于漫漫，王亚军，文才艺*

（河南农业大学植物保护学院，郑州 450002）

摘要：枯萎病是由尖孢镰刀菌（*Fusarium oxysporum*）侵染引起的一种典型的世界性土传病害，以瓜类蔬菜等枯萎病危害最为严重。近年来，由于重茬种植面积不断扩大，枯萎病已成为苦瓜生产中最重要的病害，给苦瓜生产带来了严重影响。生产上主要依赖化学药剂控制该病害，但化学防治易产生环境污染、抗药性及农药残留等问题。因此，生物防治成为防治苦瓜枯萎病的有效途径。本研究从河南省鹤壁卫贤、濮阳海通、安阳滑县、济源上官和洛阳新安等6个地区采集土壤样品13份，采用平板对峙法筛选出对苦瓜枯萎病菌有抑制作用的芽孢杆菌88株，其中，5个分离物对苦瓜枯萎病菌的拮抗效果明显（图1），盆栽防治效果75%以上，分别编号为LG16、HTA28、HTA21、LGC2和HTA23。通过形态学鉴测定和16S rDNA序列分析，5株芽孢杆菌分别为解淀粉芽孢杆菌（*Bacillus amyloliquefaciens*）、甲基营养型芽孢杆菌（*Bacillus methylotrophicus*）、枯草芽孢杆菌（*Bacillus subtilis*）、多黏芽孢杆菌（*Paenibacillus polymyxa*）和*Bacillus tequilensis*。分别利用*ituA*、*mycB*、*tasA*、*sfp*和*fenB*的特异性引物检测了分离株中伊枯草菌素、抗霉枯草菌素、抑制孢子蛋白、表面活性素、枯草菌素和丰原素等5类抗菌素合成相关基因，结果表明，LG16和HTA21含有伊枯草菌素、抑制孢子蛋白和丰原素合成相关基因；HTA28含有伊枯草菌素、抗霉枯草菌素、抑制孢子蛋白和丰原素合成相关基因；HTA23仅含有表面活性素合成相关基因；而LGC2不含有这5类抗菌素合成相关基因（表1）。目前正在开展生防菌株的定殖规律、活性物质的分离纯化及其生防机制方面的研究。

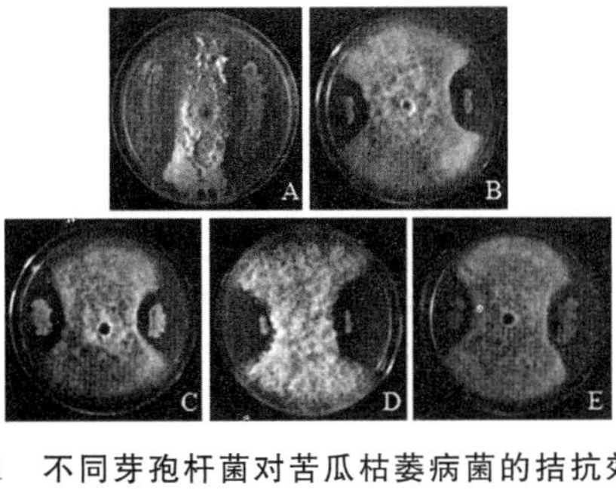

图1　不同芽孢杆菌对苦瓜枯萎病菌的拮抗效果

A：LG16；B：HTA28；C：HTA21；D：LGC2；E：HTA23

表1　不同芽孢杆菌菌株含产抗菌素基因检测结果

Strain	*ituA*	*mycB*	*tasA*	*sfp*	*fenB*
LG16	+	−	+	−	+
HTA28	+	+	+	−	+
HTA21	+	−	+	−	+
LGC2	−	−	−	−	−
HTA23	−	−	−	+	−

基金项目：公益性行业（农业）科研专项（201503110-11）

通讯作者：文才艺，博士，教授，研究方向：植物病害生物防治；E-mail：wencaiyi1965@163.com

第一作者：郭康迪，硕士研究生，植物病理学专业；E-mail：guokangdi0316@163.com。

降解无机磷的解淀粉芽孢杆菌 PHODB35 的分离筛选与鉴定

赵卫松，苏振贺，王培培，郭庆港，鹿秀云，李社增，张晓云，马平*

（河北省农林科学院植物保护研究所，
农业部华北北部作物有害生物综合治理重点实验室，
河北省农业有害生物综合防治工程技术研究中心，保定 071000）

摘要：磷是植物生长过程中不可缺少的大量元素之一。目前农业生产上大多是通过反复施加磷肥来获得高产，但随着磷肥使用量的增加，其利用率逐年降低，并且增加了农业成本。土壤中含有大量的磷，但其易与 Ca^{2+}、Al^{3+} 等离子结合形成难溶性磷酸盐，在土壤中被固定导致难以被植物吸收利用。因此，如何将土壤中的难溶性磷转化为能够被植物吸收利用的可溶性磷，筛选对作物具有促生功能的微生物，对于减少化肥的使用提供了重要的思路。本研究利用选择性培养基 NBRIP 筛选无机磷降解菌，通过钼锑抗比色法测定发酵液中可溶性磷含量达 75.74 mg/L，经 16S rRNA 和 *gyr*B 基因序列比对分析，初步鉴定 PHODB35 菌株为解淀粉芽孢杆菌（*Bacillus amyloliquefaciens*）。盆栽试验研究结果表明，接种 PHODB35 菌株发酵液后对番茄植株株高、地上部鲜重、地下部鲜重与空白对照之间均存在显著差异，增加率分别为 28.21%、22.59%、113.06%；同时经 PHODB35 菌株处理后土壤中的有效磷和番茄植株中的有效磷增长率分别为 45.08% 和 16.24%，表明 PHODB35 菌株具有较大的开发利用潜力。

关键词：无机磷降解菌；解淀粉芽孢杆菌；促生作用

基金项目：河北省自然科学基金（C2017301069）；河北省财政项目（494-0701-YBN-K9UA）

通讯作者：马平，博士，研究员，主要从事植物病害生物防治研究；E-mai：pingma88@126.com
第一作者：赵卫松，博士，助理研究员，主要从事植物病害防治和环境微生物研究；E-mai：zhaoweisong1985@163.com。

水稻细菌性穗枯病苗期病害分级标准研究

徐以华[1,2]，彭婧琦[3]，刘连盟[1]，王玲[1]，孙磊[1,2]，
梁梦琦[1]，高健[1]，黎起秦[2*]，侯雨萱[1*]，黄世文[1,2*]

([1]中国水稻研究所，杭州 310006；[2]广西大学农学院，南宁 530004；
[3]湖南农业大学植物保护学院，长沙 410000)

摘要：水稻细菌性穗枯病(bacteria panicle blight of rice，BPBR)是目前水稻生产上危害最严重的细菌性病害之一，由颖壳伯克氏菌(*Burkholderia glumae*)引起，主要危害穗期，可导致谷粒变褐、萎缩、畸形甚至不实，严重时造成80%的减产。该病害除了危害水稻穗部外，在苗期也容易发病，造成秧苗腐烂，故该病害又称为水稻苗腐病。BPBR的发生严重依赖环境条件，特别是温度和湿度。目前，随着全球气候变暖越来越利于BPBR的发生流行，但调查和诊断该病害主要还是依赖穗期特征。由于缺乏苗期病害分级标准，难以了解苗期水稻BPBR发病情况，容易忽视或降低对该病害的评估、分析和诊断。本研究旨在从苗期描述该病该特征，并制定相应的水稻细菌性穗枯病苗期病害分级标准。本研究选用50个不同抗感BPBR的水稻品种，通过浸种、催芽，然后培养水稻苗至4～5叶期，用菌液浸根，观察并描述各不同抗性品种之间的苗期病害特征，并对此病害特征差异进行分级。通过对水稻细菌性穗枯病苗期病害分级标准的制定，不仅有利于及时发现、评估和诊断该病害，还有利于更加便利的鉴定水稻抗感BPBR品种，避免了穗期接种鉴定抗性的复杂性及此过程中由气候条件不同造成发病程度的不确定性。

关键词：水稻；水稻细菌性穗枯病；苗期；病害分级标准

象耳豆根结线虫侵染红麻研究初报

王会芳[1]，芮凯[1]，曾向萍[1]，田威[2]，符美英[1]，陈绵才[1*]

（[1]海南省农业科学院植物保护研究所，海南省植物病虫害防控重点实验室，海口 571100；
[2]海南大学热带农林学院，海口 570228）

摘要：红麻（*Hibiscus cannabinu*）又名洋麻或槿麻，台湾称钟麻，为锦葵科（*Malvaceae*）木槿属（*Hibiscus*）一年生草本韧皮纤维作物，是麻纺和造纸工业的重要原料，其纤维强力大、吸湿性好、散水快，耐腐蚀磨损，应用领域十分广阔。红麻根结线虫病是一种世界性病害，在我国各红麻主产区，红麻根结线虫病也普遍发生，一般减产 20%～30%，严重者达 50%以上，甚至绝产，造成红麻生产的巨大经济损失。关于根结线虫具体的病原种类的已有不少报道。1988年，吴家琴等对采自湖南、湖北、广东、广西、浙江和河南等省（区）主产麻区的红麻根结和染病土壤按北卡罗来纳寄主反应试验法鉴定，发现所查各县（市）染病麻田均有南方根结线虫（*Meloidogyne incognita*）1 号和 2 号小种，多数麻区有花生根结线虫（*M. arenaria*）2 号小种。安徽怀远县红麻上的根结线虫的主要种类是南方根结线虫（*M. incognita*）Ⅰ 号小种和 2 号小种，长江流域和华南麻区未发现有爪哇根结线虫（*M. javanica*）分布。对 14 个红麻主产县的调查结果表明，根结线虫病病原都以南方根结线虫为主，其次为爪哇根结线虫和花生根结线虫，分别占样本总数的 76.5%、17.3%和 6.2%。以往研究表明，红麻根结线虫病原包括花生根结线虫 *M. arenaria*、爪哇根结线虫 *M. javanica* 和南方根结线虫 *M. incognita*，这些种群在田间表现为单独侵染，且在不同区域占据各自的优势地位。可见红麻根结线虫病病原表现出一定的多样性。

象耳豆根结线虫最早于 1983 年在海南儋州的象耳豆树上发现并被鉴定和命名。国内外学者报道 *M. enterolobii* 和玛雅根结线虫（*M. mayaguensis*）是同一种线虫，并已被国际上大多数学者承认。*M. enterolobii* 传播速度快，在美洲、非洲、欧洲等均有分布。2008 年，瑞士首次报道象耳豆根结线虫，经评估，认为该线虫是中欧地区作物生产的极大潜在威胁。至 2011 年，已有近 20 个国家和地区报道该线虫的发生与危害。国内，近年来 *M. enterolobii* 已在海南岛普遍分布，广东省包括深圳市、福建省等地陆续发现 *M. enterolobii* 的侵染。据调查，海南岛主栽的葫芦科蔬菜、茄果类蔬菜、南药植物和热带果树均已发现有 *M. enterolobii* 广泛寄生，显示出取代南方根结线虫而成为我国热带亚热带地区最重要根结线虫种类的趋势。

基金项目：现代农业产业技术体系（CARS-16-E10）
通讯作者：陈绵才（1959-　），男，研究员，主要从事植物病理学研究；E-mail：miancaichen@ 163. com
第一作者：王会芳（1981-　），女，副研究员，主要从事植物线虫病害方面研究。

随着我国现代设施农业的发展，温室和设施大棚等措施使环境温度的不断上升，因此 *M. enterolobii* 为害地区进一步扩大，给我国相关农作物的生产安全带来极大威胁。

为了进一步弄清红麻根结线虫的种类变化及其种群是否具有特异性，本文对采自中国南方地区湖南、福建、云南、广西、安徽、海南等省份红麻产区的 16 份红麻根结线虫种群，运用会阴花纹表型观察和同工酶分析的方法进行种类鉴定，结果发现，16 份样本中，10 份为象耳豆根结线虫（*M. enterolobii*），2 个种群为象耳豆根结线虫（*M. enterolobii*）和南方根结线虫（*M. incognita*）的混合种群，2 个种群为南方根结线虫（*M. incognita*），1 个种群为爪哇根结线虫（*M. javanica*），其余 1 个种群为花生根结线虫（*M. arenaria*）。研究结果表明，在中国南方红麻种植区，象耳豆根结线虫逐渐上升为红麻根结线虫的主要种群，呈由南向北逐渐扩散的趋势；越往南方象耳豆根结线虫侵染红麻所占比例越大。另外，本研究范围内初步表明，侵染红麻的根结线虫种类没有表现出特异性，优势种群在南方地区也呈现出一定的变化，象耳豆根结线虫逐渐取代南方根结线虫成为南方地区红麻根结线虫的优势种群。本研究中样本数量有限，更准确的数据结果有待于进一步的研究。

关键词：象耳豆根结线虫；红麻；种类

黑龙江省大豆胞囊线虫抗性资源引进与遗传改良

马兰，陈井生，李泽宇，李建英，于吉东，
吴耀坤，周长军，李肖白，田中艳

（黑龙江省农业科学院大庆分院，大庆 163316）

摘要：大豆胞囊线虫（soybean cyst nematode，SCN）病是大豆生产上的主要病害之一，其发生严重影响大豆的产量和品质。黑龙江省每年有超过1000万亩大豆遭受其危害，嫩江、依安、拜泉、大庆、安达、肇东、林甸、富裕、北安、富锦等地区尤为严重。防治大豆胞囊线虫病的主要方式是轮作、抗病育种、化学防治和生物防治等。选育抗病品种是防治大豆胞囊线虫最为经济、有效的措施。随着种植结构调整的深入推进，黑龙江省大豆面积将逐年增加，大豆胞囊线虫病亦日益严重。近年来审定推广的抗线大豆品种存在抗源基因狭窄、抗性单一（主抗3号生理小种）等问题。生产上由于连续种植大豆导致对大豆胞囊线虫形成一定选择压力，生理小种会发生毒力变异，因此对该病的防治迫切需要不同抗线虫基因来源的抗病品种。本研究从中国农业科学院、沈阳农业大学、吉林省农业科学研究院、黑龙江八一农垦大学等多家单位搜集引进抗线虫大豆资源72份，通过连续2年的适应性鉴定（抗旱、抗倒伏、生育期等）和大豆胞囊线虫病（SCN）抗性鉴定，从中筛选获得兼抗大豆胞囊线虫1号、3号生理小种的材料6份，其中，3份来自五寨黑豆改良材料后代，2份来源于PI88788，1份来自PI437654。利用其抗性优势进行有性杂交，配置杂交组合，形成一些中间骨干桥梁亲本。本研究有效拓宽了黑龙江省目前大豆抗线虫品种遗传基础，对黑龙江省抗线虫大豆育种和大豆胞囊线虫防治具有深远的意义。

关键词：大豆；大豆胞囊线虫；生理小种；抗病育种

基金项目：现代农业产业技术体系建设专项（CARS-004-CES07）；大庆市科技局指导性项目（ZD-2016-121）

第一作者：马兰（1980- ），男，助理研究员，硕士，主要从事植物线虫学和大豆抗线虫育种研究；E-mail：malan042999@163.com。

气候变暖对我国小麦条锈病菌越夏和越冬的潜在影响

赵志伟，王晓晶，王海光*

（中国农业大学植物病理学系，北京 100193）

摘要：全球气候变暖受到国际社会的广泛关注。联合国政府间气候变化专门委员会（International Panel of Climate Change，IPCC）和国内外研究人员根据不同的模式和方法对气候变暖趋势进行了研究和预测。气候变暖将对植物病害的流行规律造成重大影响。因此，在全球气候变暖背景下，开展植物病害流行规律的潜在变化对于病害管理具有重要意义。小麦条锈病菌（*Puccinia striiformis* f. sp. *tritici*）是一种对温度比较敏感的专性寄生菌，由该菌引起的小麦条锈病是我国小麦生产上的一种主要病害，该病害是一种大区流行性病害，依靠夏孢子通过高空气流远距离传播，条锈病菌的越夏和越冬对于我国小麦条锈病的周年循环和流行尤为重要。本研究基于国内外学者对到21世纪末全球和地区平均温度的预测，运用地理信息系统软件ArcGIS，利用我国历史气象数据资料和小麦条锈病发生历史资料，分别以平均年连续最热10天平均温度和平均年最冷月均温为基础，在平均温度统一升高6个不同梯度的模式下，结合温度与海拔的关系模型，运用优选的普通克里金指数模型OK(E)对我国小麦条锈病菌潜在的越夏和越冬地区进行了区划预测。进行越夏区划时，以平均年连续最热10天平均温度是否高于22～23℃作为区划温度指标，以我国小麦种植区作为区划范围；进行越冬区划时，以平均年最冷月均温是否超过－6～－7℃为区划温度指标，以我国冬麦区和冬春麦区作为区划范围。结果表明，随着气候变暖，我国适于条锈病菌越夏的区域逐渐缩减；湖北、内蒙古、山西、陕西、宁夏、云南南部和贵州在21世纪内将很可能逐渐不再适于条锈病菌越夏，陇东在全国平均温度升高2℃时将很可能不再适于条锈病菌越夏；陇南和四川北部越夏区、西藏越夏区到21世纪末极有可能仍是我国主要的条锈病菌越夏区；青海、云南和新疆到21世纪末仍有部分地区适于条锈病菌越夏。结果表明，随着气候变暖，我国主要冬麦区适于条锈病菌越冬的范围将越来越大；我国小麦条锈病菌经验越冬地理界线在21世纪内将很可能被完全打破；青海东部、新疆（除塔里木盆地外的地区）适于条锈病菌越冬地区逐渐缩小，到21世纪末将很可能成为条锈病菌不能越冬的地区，塔里木盆地将很有可能成为我国条锈病菌另外一个主要的“冬繁区”。本研究结果有助于了解气候变暖对我国小麦条锈病菌越夏和越冬的可能影响，为现在和将

基金项目：国家自然科学基金项目（31101393）

通讯作者：王海光，副教授，主要从事植物病害流行学和宏观植物病理学研究；E-mail：wanghaiguang@cau.edu.cn

第一作者：赵志伟，男，硕士研究生；E-mail：1988522zzw@163.com。

来我国小麦条锈病的综合防治工作提供了一定参考，对于其他国家小麦条锈病菌或与全球气候变暖有关的植物病虫害研究人员了解相关病虫害的未来发展趋势亦具有一定参考意义。

关键词：气候变暖；小麦条锈病；越夏；越冬；病害预测；GIS

甘肃河西地区小麦条锈病的发生在病害区域循环中的作用

孙振宇[1*]，李辉[2]，曹世勤[1]

([1]甘肃省农业科学院植物保护研究所，兰州 730070；
[2]甘肃省植保植检站，兰州 730020)

摘要：甘肃省地理环境的多样性为小麦条锈菌的越夏及越冬提供了必备条件，因此小麦条锈病在甘肃省小麦产区可完成其周年循环。但是作为甘肃省春小麦主产区的河西地区，该区域小麦条锈病的发生情况及其在病害区域循环中的作用还需明确。通过 2017 年 6 月 29 日至 7 月 3 日及 7 月 23 日至 25 日分别对甘肃省河西地区小麦条锈病发生情况的实地调查，结合对小麦条锈菌西北越夏区病害发生规律的分析，结果表明河西地区的白银市景泰县、武威市天祝县、武威市古浪县、金昌市永昌县、张掖市山丹县及张掖市民乐县的春小麦存在生育期及发病的差异，其为小麦条锈菌西北越夏区小麦收获后小麦条锈菌的越夏提供了寄主及环境条件。该结果将有助于完善对我国小麦条锈菌西北越夏区小麦条锈病流行体系的研究，是对当前气候变化条件下我国小麦条锈病发生流行新情况的探索。

关键词：甘肃省；河西地区；小麦条锈病；病害循环

基金项目：国家自然科学基金项目(31660498，31301605)

通讯作者：孙振宇，博士，副研究员，主要从事植物病害流行学研究；E-mail：szy20020815@163.com

电子商务影响植物病毒病的远距离传播

赵航海[#]，卢春晓[#]，张雪琳，韩成贵，王颖[*]

（中国农业大学植物病理学系，北京 100193）

摘要：植物病毒是一种分子寄生物，在植物上是仅次于真菌的第二大病原，给农业带来严重危害。在自然界中，病毒主要依靠介体和非介体进行远距离传播，但由于近年来电子商务的迅速发展，在极大方便人们生活的同时也为病毒的传播提供了新途径。本实验室从2个电商平台上随机挑选5家网店购买了18份百合鳞茎，利用RT-PCR的方法对其进行了8种病毒病的检测，并对阳性样品进行测序。结果显示5家网店出售种球的带毒率为3.3%～12.1%不等；21.21%的鳞茎携带百合斑驳病毒（*Lily mottle virus*，LMoV），12.12%的鳞茎携带车前草花叶病毒（*Plantago Asiatic mosaic virus*，PLAMV），6.06%的样品中同时检测到两种病毒。PLAMV是目前荷兰百合的主要病毒病之一，我国于2015年首次在进口荷兰百合中检出PLAMV，而如今电子商务更是将此病毒的感染范围进一步扩大，可能给我国花卉产业带来潜在威胁。同时实验室也对电商平台上售卖的茄科和葫芦科种子分别进行了番茄斑萎病毒和黄瓜绿斑驳花叶病的检测，结果未检测到相应病毒，说明网络平台上销售的种子还未存在病毒病扩散的隐患。本研究表明，电商平台中售卖的百合鳞茎大部分为塑料简易包装，没有正规商品化，有一部分携带病毒，暗示着通过网络平台售卖的没有正规商品化的种子、种苗等会存在带毒情况，尤其是对一些检疫病毒，目前对电子物流中植物材料的检疫还没有很严格的规定，对我国控制有害生物扩散的工作是十分不利的。

基金项目：中国农业大学植物保护学院本科生URP项目

通讯作者：王颖，副教授，主要从事植物病毒学研究；E-mail：yingwang@cau.edu.cn

共同第一作者：赵航海，硕士生，主要从事植物病毒与寄主的分子互作研究；E-mail：hanghaizhao@cau.edu.cn

卢春晓，本科生；E-mail：1401081021@cau.edu.cn。

2 株辣椒疫霉生防细菌的鉴定与生防潜力研究

汪涛，赵伟，迟元凯，戚仁德*

（安徽省农业科学院植物保护与农产品质量安全研究所，合肥 230031）

摘要：辣椒疫病是辣椒生产中危害最严重的病害之一，常引起移栽时的缺棵断垄和挂果期枯萎，对辣椒安全生长造成重大威胁。以辣椒疫霉为指示菌，通过平板对峙培养，从大别山区连续种植 30 年以上的蔬菜地土壤中筛选到 2 株细菌 WT3 和 TC35，通过形态特征、生理生化特性以及 16SrDNA 和 gyrB 基因序列比对，将这两个菌株鉴定为多粘类芽孢杆菌；田间药效试验结果表明，多粘类芽孢杆菌 WT3 和 TC35 的发酵液对辣椒疫病均具有良好的防控效果，其中 WT3 的防效达到了 85.0%，TC35 对辣椒疫病的防效达到了 92.1%，对照药剂烯酰吗啉的防效为 86.4%。多粘类芽孢杆菌 WT3 和 TC35 对辣椒疫病具有良好的生防效果，可作为防治辣椒疫病的生物农药进行研究与开发。

关键词：辣椒疫病；生物防治；多粘类芽孢杆菌；鉴定；田间药效试验

基金项目：安徽省农业科学院学科建设(15A1107)

通讯作者：戚仁德，研究员，从事蔬菜病害及其防控技术研究；E-mail: rende7@126.com

第一作者：汪涛，男，安徽岳西人，助理研究员，从事蔬菜病害及其防控技术研究；E-mail: chinawtty@163.com。

产酶溶杆菌 OH11 中信号分子 4-HBA 参与抗菌物质 HSAF 生物合成的机制研究

苏振贺[1],王培培[1],赵卫松[1],张晓云[1],鹿秀云[1],李社增[1],郭庆港[1],钱国良[3],刘凤权[2],马平[1*]

([1]河北省农林科学院植物保护研究所,
河北省农业有害生物综合防治工程技术研究中心,
农业部华北北部作物有害生物综合治理重点实验室,保定 071001;
[2]江苏省农业科学院植物保护研究所,南京 210014;
[3]南京农业大学植物保护学院,农业部华东作物有害生物综合治理重点实验室,南京 210095)

摘要:产酶溶杆菌 OH11 是具有生防潜力的生防细菌,该菌产生拮抗丝状真菌的热稳定因子 HSAF (heat stable antifungal factor),具有十分广阔的潜在应用价值。DF (diffusible factor) 是细菌中一种重要的群体感应系统,DF 包含 4-羟基苯甲酸(4-HBA)和 3-羟基苯甲酸(3-HBA),前期研究发现 DF 参与调控 HSAF 的生物合成,但是具体机制未知。通过生物信息学预测、遗传途径证实莽草酸途径参与 DF 及 HSAF 的生物合成,生化遗传数据表明 LenB2 是产生 DF 的关键酶,该酶以莽草酸途径终产物分支酸为底物产生 3-HBA 和 4-HBA,对一系列相关突变体体外添加微量 DF,0.5 μM 4-HBA 即能恢复相关突变体 HSAF 产量,揭示 4-HBA 作为一种信号分子正向调控了 HSAF 的生物合成。生物信息学预测 $LysR_{Le}$是 4-HBA 的受体,并且 $\Delta lysR_{Le}$不产生 HSAF,而 lysRLe 正向调控 HSAF 的机制是 $LysR_{Le}$ 转录调控因子结合在 HSAF 生物合成关键基因 *lafB* 启动子从而调控 HSAF 的生物合成。微量热涌动实验表明 $LysR_{Le}$能够直接结合 4-HBA,通过 EMSA 实验证实,4-HBA 能够增强 $LysR_{Le}$与 HSAF 生物合成关键基因 *lafB* 启动子的结合,揭示 4-HBA 通过作用于 $LysR_{Le}$来正向调控 HSAF 的生物合成。该机理的研究为遗传改良获得高产 HSAF 菌株奠定理论基础。

关键词:产酶溶杆菌;4-HBA;HSAF

通讯作者:马平(1964-),男,博士,研究员,主要从事植物病害生物防治研究;E-mail:pingma88@126.com
第一作者:苏振贺(1986-),男,博士,助理研究员,主要从事植物病害生物防治研究;E-mail:suzhenhe0202@126.com。

解淀粉芽孢杆菌 Jt84 防治水稻稻瘟病机制初探

张荣胜，王法国，齐中强，杜艳，刘永锋*

（江苏省农业科学院植物保护研究所，南京 210014）

摘要：生防芽孢杆菌防治植物病害主要通过竞争营养与空间、产生具有抑菌和溶菌活性代谢产物、诱导植株产生系统抗病性等多方面协同作用，同时生防芽孢杆菌发挥作用的前提是在植株上有效定殖。我们研究室前期筛选获得一株解淀粉芽孢杆菌 Jt84，通过喷雾干燥法制备成干悬浮剂，田间试验表明该菌对水稻穗颈瘟具有良好的防治效果，防效为79.2%，与化学药剂三环唑防效相当。为了明确其对稻瘟病的防治机制，本研究围绕解淀粉芽孢杆菌 Jt84 产生抗菌物质的种类和定殖规律开展相关工作，为深入研究解淀粉芽孢杆菌 Jt84 防治稻瘟病提供理论基础，同时为解淀粉芽孢杆菌 Jt84 的推广与应用提供技术支撑。通过4对特异性引物 PCR 检测生防菌 Jt84 中相关抗菌物质合成基因，表明解淀粉芽孢杆菌基因组中含有 *sfp*、*fenB*、*ituA* 或 *bamA*、*mycB* 基因。平板对峙试验结果也显示，解淀粉芽孢杆菌 Jt84 干悬浮剂提取的脂肽类物质对稻瘟病菌拮抗带宽为 9.8 mm；利用液质联用方法检测，发现干悬浮剂的甲醇萃取液中含有 surfactins、fengycins 和 iturins 3 大类家簇脂肽类抗生素。前人研究表明，iturin 类抗菌物质在防治植物真菌病害中起着关键作用，我们构建了 *ituA* 和 *mycB* 双交换载体，经化学转化获得 *ituA*、*mycB* 敲除突变株，平板抑菌试验显示突变株对稻瘟病菌拮抗能力降低约 50%。生防菌 Jt84 定殖动态结果显示：在水稻三叶期时，喷施初始菌量为 5.5×10^8 cfu/mL 的 JT84gfp^{rif} 在水稻叶片上；喷施后 2 h 取样检测，叶片上的菌量为 4×10^6 个/cm²，喷施后 7～11 d，芽孢杆菌菌量维持稳定（回收菌量为 10^4 个/cm² 左右），喷施后 30 d 仍然可以回收到菌量为 22 个/cm²，说明解淀粉芽孢杆菌 Jt84 能够有效定殖于水稻叶片上。建议田间使用解淀粉芽孢杆菌 Jt84 干悬浮剂防治稻瘟病时，药剂喷施后 10 d 左右，再喷施一次生防菌确保对稻瘟病的有效防治。

基金项目：国家重点研发计划（2016YFD0300706）；江苏省自主创新资金项目 CX15(1054)

通讯作者：刘永锋，研究员，主要从事植物病害致病机制及其防控技术研究；Tel：025-84391002，E-mail：liuyf@jaas.ac.cn

第一作者：张荣胜，男，江苏泗洪人，博士，副研究员，主要从事水稻病害生物防治及其应用技术研究；E-mail：r_szhang@163.com。

巨大芽孢杆菌 Sneb207 诱导大豆抗胞囊线虫的差异转录组学分析

周园园,朱晓峰,王媛媛,刘晓宇,范海燕,段玉玺,陈立杰*

(沈阳农业大学植物保护学院,沈阳 110866)

摘要:大豆胞囊线虫病是大豆生产上危害最为严重的根部病害。世界范围内,几乎所有的大豆产区均发生该病害。利用生防菌控制大豆胞囊线虫病是一种持续有效的措施。研究较多的生物因子包括食线虫真菌、细菌、放线菌和杀虫植物等。其中芽孢杆菌是一类防控土传病害效果很好的生防因子,有些以生物农药或生物菌肥的形式在农业生产中推广使用。本实验室前期筛选到一株巨大芽孢杆菌 Sneb207 可提高大豆对大豆胞囊线虫的抗性,在包衣和不包衣大豆种子的条件下,对比接种和不接种线虫 J2 的大豆根系差异转录组,将筛选的差异表达基因与 KEGG 数据库对比结果,并通过 qRT-PCR 将筛选到的差异基因进行进一步验证,发现经生防菌包衣处理后,植物激素信号转导通路中注释 50 个差异基因;植物病原物互作通路中,注释到 27 个差异基因;淀粉和蔗糖通路中,注释到 26 个差异基因;苯丙烷类生物合成通路中,注释到 32 个差异基因,而未包衣处理注释到的差异基因少于包衣处理。因此,经生防菌包衣处理后,可能这几个通路起主要的抗线虫作用。根据转录组结果筛选到的 31 个抗病相关基因并进行 qRT-PCR 验证,结果表明 qRT-PCR 中基因表达的数据与 RNA-Seq 结果一致。本研究初步从转录组学水平上明确了巨大芽孢杆菌 Sneb207 诱导大豆产生抗线虫的作用机制与植物激素信号转导、植物病原物互作、淀粉和蔗糖通路及苯丙烷类生物合成通路相关性最大。

关键词:巨大芽孢杆菌 Sneb207;大豆;大豆胞囊线虫;RNA-Seq 转录组分析

基金项目:农业部公益性行业科研专项(201503114);国家自然科学基金重点项目(31330063);国家自然科学基金面上项目(31171569,31571985);现代农业产业技术体系建设专项大豆线虫防控岗位 CARS-04-PS13

通讯作者:陈立杰,博士研究生,教授,主要从事植物线虫学研究;E-mail:chenlijie0210@163.com

第一作者:周园园,博士研究生,主要从事植物线虫学研究;E-mail:zhouyuanyuan6616@163.com。

枯草芽孢杆菌 BAB-1 代谢产物对黄瓜白粉病的防治作用

张晓云,鹿秀云,郭庆港,李社增,王培培,赵卫松,马平*

(河北省农林科学院植物保护研究所/河北省农业有害生物综合防治工程技术研究中心/农业部华北北部作物有害生物综合治理重点实验室,保定 071000)

摘要:枯草芽孢杆菌 BAB-1 是本实验室筛选出的一株对黄瓜白粉菌及番茄灰霉菌等气传病原菌具有较强抑制作用的生防菌。为明确菌株 BAB-1 的抑菌机理,本试验采用子叶喷雾法探究菌株 BAB-1 代谢产物不同成分对黄瓜白粉病的防治效果;通过灌根法研究该菌株代谢产物不同成分的诱导抗病能力。结果表明,菌株 BAB-1 的最佳发酵时间为 44 h,此时发酵液(100 倍稀释)对黄瓜白粉病的防效高达 91.56%;发酵上清液和菌体悬浮液稀释 100 倍后的防效分别为 83.47%和 28.48%,说明发酵上清液在黄瓜白粉病的防治中发挥了主要作用;发酵上清液中的蛋白(100 倍稀释)对黄瓜白粉病的防效为 69.96%,而脂肽(100 倍稀释)的防效仅为 16.65%;诱导抗性试验结果表明菌株 BAB-1 发酵上清液能够诱导黄瓜产生抗病性,并且上清液中的蛋白是诱导黄瓜对白粉病产生抗性的主要成分,其诱导防效为 19.2%,而脂肽不具诱导抗性作用。

关键词:枯草芽孢杆菌;蛋白;脂肽;防病机理

基金项目:国家重点研发计划-高效细菌杀菌剂的研制与示范(2017YFD0201101)

通讯作者:马平,博士,研究员,主要从事植物病害生物防治研究;E-mail:pingma88@126. com

第一作者:张晓云,硕士,助理研究员,主要从事植物病害生物防治研究;E-mail:zxy_zxl@163.com。

生防菌盾壳霉胞外丝氨酸蛋白酶 CmSp1 的功能研究

王永春，杨龙*，李国庆

（华中农业大学植物科学技术学院，武汉 430070）

摘要：核盘菌[*Sclerotinia sclerotiorum*（Lib.）de Bary]是一种重要的植物病原真菌，可以引起多种作物的菌核病。重寄生真菌盾壳霉（*Coniothyriumminntans*）对核盘菌专性寄生，对作物菌核病的生物防治有着广阔的应用前景。盾壳霉在对核盘菌重寄生过程中会分泌大量的胞外丝氨酸蛋白酶，但对其分泌的丝氨酸蛋白酶功能及酶学特性等仍不清楚。我们采用真核核表达系统，对盾壳霉胞外丝氨酸蛋白酶基因 *CmSp1* 进行了真核表达。首先通过 PCR 扩增 *CmSp1* 基因的 cDNA 序列，构建真核表达载体 pPIC9-*CmSp1-6His*。电击转入毕赤酵母 GS115，获得了酵母转化子，采用福林酚法测定了酵母转化子蛋白酶诱导表达的时间动态。*CmSp1* 蛋白酶酵母转化子在培养 60 h 时，达到最高为 12.39 U/mL。接着对蛋白酶 *CmSp1* 酵母转化子产生的蛋白酶进行了硫酸铵沉淀及镍柱纯化，SDS-PAGE 分析纯化效果良好，蛋白大小为 35 ku。采用 His 标签做抗体，对 CmSp1 蛋白酶进行 western blot 分析，杂交得到一条约 35 ku 的蛋白条带，与 SDS-PAGE 分析结果吻合。证明该纯品即是带 His 标签的 CmSp1 蛋白酶纯品。明胶平板试验证实该 CmSp1 蛋白酶纯品具有蛋白酶活性。进一步通过福林酚法对 CmSp1 蛋白酶的酶学特性进行了测定，CmSp1 反应最适温度为 40℃，最适反应 pH 为 7。对丝氨酸蛋白酶抑制剂 PMSF 敏感。用浓度为 5 mmol/L 的 Mg^{2+}、Fe^{2+}、Mn^{2+}、Li^{+}、Ca^{2+}、Zn^{2+}、K^{+}、Cu^{2+} 离子处理 CmSp1，其中 Mn^{2+}、Ca^{2+} 和 Cu^{2+} 离子处理后，蛋白酶活性分别是原来的 1.45 倍、1.14 倍和 1.16 倍。最后分析了蛋白酶 CmSp1 的生防活性。采用牛津杯法测定了 CmSp1（酶活：0.926 U/mL）对核盘菌和灰葡萄孢菌菌丝生长的抑制作用，结果显示，CmSp1 对核盘菌及灰霉菌的菌丝生长有显著的抑制作用。上述结果说明：胞外丝氨酸蛋白酶 CmSp1 具有生物活性。对其生物学功能进行解析将有助于揭示其在盾壳霉重寄生中的作用。

基金项目：国家自然科学基金（31471813）

通讯作者：杨龙，副教授，研究方向为植物病害生物防治；E-mail：yanglong@mail. hzau. edu. cn

第一作者：王永春，博士研究生；E-mail：yongchun@webmail. hzau. edu. cn。

醉蝶花对黄瓜枯萎病菌的抑菌活性及成分分析

高继杨，刘大伟，刘行风，王利锋，张艳菊*

（东北农业大学农学院植物保护专业，哈尔滨 150030）

摘要：由无性菌类尖孢镰孢菌（*Fusarium oxysporum*）侵染引起的枯萎病是严重威胁黄瓜生产的土传病害。采用化学药剂对土壤进行消毒和熏蒸处理是防治黄瓜枯萎病的常用且有效方法。然而化学药剂不仅对土壤造成污染，而且也给人类身体健康带来很大危害。因此，寻找环保有效、可替代化学药剂的生物熏蒸剂成为当务之急。本研究采用菌丝生长速率法研究了醉蝶花茎叶粉碎物对黄瓜枯萎病菌的熏蒸作用，应用 GC/MS 检测了醉蝶花挥发性物质的有效成分，并对各成分的抑菌活性进行了研究。结果表明，0.1 g、0.2 g、0.3 g 和 0.4 g 4 个剂量的醉蝶花粉碎物对黄瓜枯萎病菌均有抑制作用，其中 0.3 g 和 0.4 g 抑菌率可达 100%。对醉蝶花提取物采用 GC/MS 并利用 N IST05 标准质谱库-计算机联机检索，共鉴定出了 21 种化合物，包括 13 种杂环，1 种醇，1 种色烯，1 种酯，1 种酸以及 6 种长链烷烃。其中含量最多的是牡丹酚（22.60%），其次为 1，1，4a-三甲基-3，4，4a，5，6，7-六氢-2-(1H)-萘酮（12.56%），2，5，5，8a-四甲基-3，4，4a，5，6，8a-六氢-2H-色烯（8.00%），9，12，15-十八碳三烯酸甲酯（6.78%），脂酸（5.04%），沉香醇（3.74%），3-乙基-5-(乙基丁基)-十八烷（3.46%），植酮（3.29%），十四烷基环氧乙烷（3.19%），9-乙基-十七烷（3.17%），三十五碳-17-烯（3.16%），二十四烷（3.13%），7-羟基-四甲基-7，8-二氢环庚并吡喃[d，e]萘（3.10%），7，12a-二甲基-1，2，3，4，4a，11，12，12a-八氢化屈（2.93%），十七烷（2.89%），乙基环戊烯醇酮（2.69%），茶香螺烷（2.51%），二十八烷（2.09%），8-甲基-5(4-戊烯基)八氢吲哚嗪（1.95%），2，6-二甲基-1-环己烯-乙酸酯（1.92%），4-羟基-β-紫罗兰酮（1.81%）。对其中 8 种成分的熏蒸效果进行研究，结果表明牡丹酚、芳樟醇和茶香螺烷对黄瓜枯萎病菌有抑制作用，抑菌率分别为 73.8%、75.9%和 80.4%，而十四烷基环氧乙烷、二十四烷、十七烷、乙基环戊烯醇酮和二十八烷等对黄瓜枯萎病菌无抑制作用。研究结果对于今后醉蝶花在生物熏蒸方面的进一步应用具有重要意义。

关键词：醉蝶花；黄瓜枯萎病；牡丹酚；芳樟醇；茶香螺烷

盾壳霉响应草酸的转录组分析

徐玉平，杨龙*，李国庆

（华中农业大学植物科学技术学院，武汉 430070）

摘要：盾壳霉（*Coniothyrium minitans*）是作物菌核病病原核盘菌（*Sclerotinia sclerotiorum*）的重寄生菌，是一种重要的生防真菌。在盾壳霉与核盘菌互作早期，盾壳霉感应核盘菌分泌草酸所营造的酸性 pH 信号来调控相关基因的表达，从而发挥降解草酸作用和抗生作用；在互作后期，由于草酸被降解后 pH 升高，盾壳霉通过 CmpacC 信号途径来调控重寄生相关基因的表达，从而发挥重寄生作用。但是，目前对盾壳霉感应草酸信号的分子机制及其生防功能尚不清楚。本文通过高通量测序技术对盾壳霉接触草酸前与接触草酸后的 20 min、1 h、3 h、6 h、9 h 的转录谱进行了转录组测序和功能注释分析。测序结果共获得了 51813 条Unigenes，通过与 NR、NT、KO 等 7 大数据库比对，其中共有 43028 条 Unigenes 的功能被注释，占 83.04%。相对于草酸处理前，483 个差异表达基因（differeitially expressed genes, DEGs）在草酸处理后表达显著上升，而 586 个 DEGs 在草酸处理后表达显著下降。上调表达的基因在 KEGG 数据库中主要富集在氧化磷酸化、丁酸代谢通路上，包括细胞色素氧化酶复合体（COXn）、还原型辅酶脱氢酶复合体（NDn）以及乙酰乙酰辅酶 A 合成酶（AACS）等，表明盾壳霉响应草酸涉及细胞内的氧化还原稳态以及氨基酸的代谢，而草酸作为一种强还原剂，确实具备影响生物体内氧还稳态和氨基酸代谢的能力。另一方面，上调表达的基因在 GO 数据库中主要富集在铜离子跨膜运输和电子传递链上。铜离子作为多种氧化酶的辅助因子参与重要的生物反应，从侧面也佐证了盾壳霉响应草酸时首要问题是维持体内氧化还原稳态。下调表达的基因在 KEGG 数据库中主要富集在 MAPK 信号通路上，包括经典蛋白激酶 C（CPKC）、渗透压的双组分系统（SLN1），表明草酸引起的细胞胁迫会激活 MAPK 信号转导通路。此外，在转录组测序结果中盾壳霉草酸脱羧酶基因 *CmOxdc*1 正向响应草酸以及 pH 响应转录因子 *CmpacC* 负向响应草酸，这与本实验室前期研究结果一致。本结果为深入研究盾壳霉响应草酸的分子机制提供了实验数据，有助于进一步明确盾壳霉与核盘菌互作分子机制，对挖掘盾壳霉生防潜力提供新思路。

关键词：盾壳霉（*Coniothyrium minitans*）；核盘菌（*Sclerotinia sclerotiorum*）；草酸；转录组测序

基金项目：国家自然科学基金面上项目（31672073）

通讯作者：杨龙，副教授，研究方向为植物病害生物防治；E-mail：yanglong@mail. hzau. edu. cn

第一作者：徐玉平，博士研究生，研究方向为植物病理学；E-mail：yupingxu@webmail. hzau. edu. cn。

小RNA在生物防治中的作用研究

汪顺娥，郭坚华，赵弘巍，牛冬冬*

（南京农业大学植物保护学院植物病理系，南京 210095）

摘要：蜡质芽孢杆菌 AR156 是一株根围促生细菌，它能够通过诱导植物水杨酸和茉莉酸信号通路产生对病原细菌和真菌的系统抗病性。研究中我们发现在生防细菌 AR156 处理后接种病原菌，一些小 RNA 被显著诱导或抑制表达。其中 miR825 和 miR825*（miR825 的互补链）的表达水平下降，进一步分析发现 miR825 和 miR825* 的作用靶标是 NB-LRR 家族基因，从而明确了 miR825 和 miR825* 通过调控 NB-LRR 家族基因，参与生防细菌 AR156 诱导植物免疫应答的调控机制。同时，我们还发现当 AR156 菌液注射拟南芥叶片时，能够激发类似 flg22 产生的 PTI 反应。AR156 激活的 PTI 反应包括蛋白激酶 MPK3/6 的磷酸化，防御相关基因 *PR1*，*FRK1*，*WRKY22* 和 *WRKY29* 的诱导表达，活性氧爆发和胼胝质积累。AR156 预处理能显著抑制 Pst 在 NahG 转基因植株以及 *jar1*、*etr1*、*ein2*、*npr1* 和 *fls2* 突变体中的增殖，这说明 AR156 诱导的 PTI 反应有可能不需要水杨酸，茉莉酸，乙烯信号和 flg22 的受体激酶 FLS2。此外，AR156 能调控多个 AGO1 结合的 miRNA 的差异表达。全基因转录组分析表明 AR156 和 flg22 激活了植物类似的转录水平，共同调控 117 个基因的差异表达，其中通过 qRT-PCR 分析证实了 16 个差异表达基因。综合以上研究结果，小 RNA 在生防细菌诱导抗病中具有重要的调控作用。

关键词：拟南芥；AR156；flg22；PTI

通讯作者：牛冬冬，副教授，主要从事小 RNA-生防-植物互作的分子机制研究；E-mail：ddniu@njau.edu.cn

第一作者：汪顺娥，在读博士生，研究方向为植物诱导抗性机理研究。

成年态柑橘接穗组培-冷冻法脱除黄龙病-病毒病复合病原研究

程保平[1]，赵弘巍[2]，彭埃天[1*]，宋晓兵[1]，崔一平[1]，凌金锋[1]

([1]广东省农科院植物保护研究所/广东省植物保护新技术重点实验室，广州 510640；
[2]南京农业大学植物保护学院，南京 210095)

摘要：本研究提供了一种利用组培-冷冻结合的方法脱除成年态柑橘接穗中的黄龙病-病毒病复合病原的技术。首先在采穗树上定期喷施利巴维林与膦甲酸的混合液，1个月后，采集健壮枝条，浸于氨苄青霉素与苄基青霉素的混合液中进行初步脱毒处理，然后置于发芽培养基诱导芽条产生，待接穗发芽后，切取芽尖进行冷冻脱毒，恢复培养后取茎尖微芽嫁接于健康试管砧木苗上，最后移苗至温室并进行病原菌检测。结果表明：该方法对黄龙病、衰退病和裂皮病病原的脱毒率，在甜橙上分别为100%、93.33%和90%，在沙田柚上分别为100%、100%和93.75%；该脱毒方法显著好于传统的热处理脱毒方法，后者对黄龙病、衰退病和裂皮病病原的脱毒率，在甜橙上分别为85.29%、67.65%和26.47%，在沙田柚上分别为77.78%、64.52%和32.26%。

关键词：柑橘黄龙病；柑橘衰退病；柑橘裂皮病；柑橘脱毒

基金项目：广东省重大科技计划专项（2014B020203003）；广东省自然科学基金（2014A030310198）；广东省现代农业产业技术体系建设专项（2016LM1077）；广东省科技计划项目（广东省农作物病虫害绿色防控技术研究开发中心建设）；广东省农业标准化项目（粤财农[2016]310号）；广东省科技计划重点实验室项目（2014B030301053）

通讯作者：彭埃天（1962-　），男，研究员；E-mail：pengait@163.com
第一作者：程保平（1982-　），男，助理研究员；E-mail：1093449912@qq.com。

蚯蚓粪对黄瓜枯萎病及根际土壤微生物多样性的影响

白玲玲，刘欣欣，张艳菊*，刘大伟，王利锋，王春龙

（东北农业大学农学院，哈尔滨 150030）

摘要：黄瓜枯萎病是一种土传病害，对黄瓜的生产造成严重影响，甚至带来严重的经济损失。近些年，蚯蚓粪作为新型生物肥料及对土传病害具有一定的作用而备受关注。本文研究了施用蚯蚓粪对黄瓜枯萎病的预防效果及对土壤环境及黄瓜生长、黄瓜品质和产量的影响；基于 Illumina HiSeq 测序研究施用蚯蚓粪对黄瓜根际土壤微生物多样性的影响。研究结果如下：苗期盆栽试验研究了 0、25%、50%、75%、100% 5 个不同配比（体积比）蚯蚓粪对黄瓜苗期生长及枯萎病的影响。结果表明施用蚯蚓粪对黄瓜的出苗率、茎粗、叶宽、叶片数及枯萎病具有显著性差异，当蚯蚓粪比例为 25%时，黄瓜的出苗率、茎粗、叶宽和叶片数最高，且对黄瓜枯萎病预防效果最佳，达 74.1%。田间栽培试验研究了不同配比蚯蚓粪对土壤环境、黄瓜枯萎病的预防效果及黄瓜品质、产量的影响，结果表明蚯蚓粪可改善土壤的理化性质，提高土壤酶活性；对预防黄瓜枯萎病具有良好的效果，25%蚯蚓粪处理预防效果可达 58.8%；添加蚯蚓粪可促进黄瓜植株的生长，随着蚯蚓粪添加比例的提高，黄瓜株高、茎粗及叶面积、维生素 C、可溶性糖、可溶性蛋白等含量及单株产量呈先增大后减小的趋势，以蚯蚓粪比例为 25%时效果最佳，且此时黄瓜的硝酸盐含量最低。利用 Illumina MiSeq 测序平台研究施用 25%蚯蚓粪对黄瓜根际土壤细菌多样性的影响，添加蚯蚓粪可改变处理中菌群的丰度，其中变化最显著的细菌门为 Proteobacteria（变形菌门）和 Bacteroidetes（拟杆菌门）；变化最显著的细菌科为 Piscirickettsiaceae（鱼立克次体）、Cytophagaceae（噬纤维细菌科）和 Peptostreptococcaceae（消化链球菌科）；变化最显著的细菌属为 *Luteimonas*（单胞菌属），且认为 *Luteimonas* 对防治黄瓜枯萎病具有一定的效果。从 OTU 数目、Alpha 多样性、Beta 多样性、veen 图及属水平物种丰度聚类热图可知，蚯蚓粪影响并改变了土壤中细菌种类。利用 Illumina MiSeq 测序平台研究施用 25%蚯蚓粪对黄瓜根际土壤真菌多样性的影响，结果表明：添加蚯蚓粪可改变处理中菌群的丰度，其中变化最显著的真菌门为 Ascomycota（子囊菌门）；变化最显著的真菌科为 Un-s-Sordariales sp；变化最显著的真菌属为 Un-s-Sordariales sp（粪壳菌目中的一个属）。并推断 Un-s-Sordariales sp 对黄瓜枯萎病的防治具有一定的作用。从 OTU 数目、Alpha 多样性、Beta 多样性、veen 图及属水平物种丰度聚类热图可知，蚯蚓粪可影响并改变土壤中真菌种类。

关键词：黄瓜枯萎病；蚯蚓粪；黄瓜生长；根际微生物多样性

蚯蚓粪对番茄根结线虫病的控制效果评价

刘大伟[1]，韩文昊[2]，梁芷健[1]，陶磊[1]，刘霆[3]

([1]东北农业大学农学院植物保护系，哈尔滨 150030；
[2]浙江大学农业与生物技术学院昆虫科学研究所，杭州 310058；
[3]北京市农林科学院植物保护环境保护研究所，北京 100097)

摘要：通过室内生测和盆栽试验，研究了蚯蚓粪及其浸提液对南方根结线虫的防治作用。研究结果表明：蚯蚓粪浸提液对南方根结线虫卵囊和卵的孵化表现出明显抑制作用，并且随着浸提液浓度的增加而提高。对二龄幼虫具有明显的击倒和致死作用，该作用也随着浓度和处理时间的增加而增强。盆栽试验表明：栽培基质混入蚯蚓粪对番茄根结线虫病起到一定的控制作用，番茄根部产生的根结数量降低，但防效并非随着蚯蚓粪比例的增加而提高。当蚯蚓粪占栽培基质的比例为75%时，对根结线虫病的防效可达70%以上，但对番茄根系的生长产生了不良影响，植株生长受抑，地上部鲜重降低，而当蚯蚓粪比例为25%时，番茄的生长情况最好，同时对根结线虫病的防治效果显著，达到了36.3%。

关键词：蚯蚓粪；浸提液；番茄根结线虫；防治效果

基金项目：东北农业大学“青年才俊”项目(16QC01)；黑龙江省普通本科高等学校青年创新人才培养计划
第一作者：刘大伟(1983-　)，男，副教授，博士，研究方向为植物病原线虫学；E-mail：liudawei353@163.com。

苜蓿镰孢菌根腐病防治药剂筛选

孔前前，阮柳，秦丰，马占鸿，王海光*

（中国农业大学植物病理学系，北京 100193）

摘要：苜蓿是一种世界广泛栽培的多年生豆科牧草，不仅含有丰富的蛋白质与维生素，具有“牧草之王”之称，而且可以改良土壤特性，防止水土流失。随着畜牧业对苜蓿需求量的增加，苜蓿种植面积逐渐增长，苜蓿病害日益引起关注，其中苜蓿根腐病一旦大面积发生会严重降低苜蓿产量，甚至需要重新建植，并且引起该病害的镰孢菌（*Fusarium* spp.）可能产生毒素，从而影响以苜蓿为饲料的牲畜健康，该病害已成为限制苜蓿产业发展的一个重要因素。苜蓿根腐病病原种类多，镰孢菌是引起该病害的主要病原菌。本研究以从河北部分苜蓿种植地采集的苜蓿根腐病样品中分离获得的4个致病性较强的菌株QD13-2（*F. oxysporum*）、QZ3（*F. equiseti*）、N12-1（*F. acuminatum*）和N1-2（*F. redolens*）作为供试菌株，进行了防治药剂筛选，旨在为苜蓿根腐病的有效防控提供一定依据。

选取98%多菌灵、96.4%福美双、98%嘧菌酯、98%咯菌腈、98.2%恶霉灵、96.3%苯醚甲环唑和96%甲基硫菌灵7种杀菌剂原药，在PDA培养基上采用含毒介质法对4个菌株进行室内毒力测定。采用垂直十字交叉法测量菌落直径，利用抑制中浓度（EC50）比较各药剂的毒力大小。结果表明，7种杀菌剂对4个菌株的菌丝生长均有一定抑制作用，并且抑制效果随药剂浓度的减小而降低，98%多菌灵、96.4%福美双、98%嘧菌酯、98%咯菌腈、98.2%恶霉灵、96.3%苯醚甲环唑、96%甲基硫菌灵对菌株QD13-2的EC_{50}分别为2.06 μg/mL、185.27 μg/mL、119.85 μg/mL、1.04 μg/mL、249.22 μg/mL、13.79 μg/mL、46.15 μg/mL，对菌株QZ3的EC_{50}分别为4.47 μg/mL、295.85 μg/mL、54.91 μg/mL、0.59 μg/mL、270.02 μg/mL、68.91 μg/mL、6.43 μg/mL，对菌株N12-1的EC_{50}分别为1.91 μg/mL、1.57 μg/mL、111.85 μg/mL、0.72 μg/mL、20.01 μg/mL、105.44 μg/mL、3.37 μg/mL，对菌株N1-2的EC_{50}分别为16.52 μg/mL、9.62 μg/mL、20.20 μg/mL、0.47 μg/mL、298.90 μg/mL、27.38 μg/mL、113.96 μg/mL。咯菌腈对4个菌株菌丝生长的抑制效果最好，其次是多菌灵和甲基硫菌灵。

将经过灭菌处理的中苜1号苜蓿种子播种到装有无菌土（营养土∶蛭石＝1∶1）的塑料花盆中，温室育苗45 d后，利用供试菌株麦粒接种体进行人工接种。接种5 d后灌施供试杀菌剂50%咯菌腈WP、50%多菌灵WP和70%甲基硫菌灵WP，施药7 d后，调查发病情况，计算药

基金项目：公益性行业（农业）科研专项经费项目（201303057）

通讯作者：王海光，副教授，主要从事植物病害流行学和宏观植物病理学研究；E-mail：wanghaiguang@cau.edu.cn

第一作者：孔前前，女，硕士研究生；E-mail：1439918864@qq.com。

剂防治效果。结果表明，对菌株 QD13-2 引起的根腐病，50%咯菌腈 WP、50%多菌灵 WP、70%甲基硫菌灵 WP 的防效分别为 22.97%、51.35%、35.14%；对菌株 QZ3 引起的根腐病，3 种药剂的防效分别为 9.46%、12.16%、12.16%；对菌株 N12-1 引起的根腐病，3 种药剂的防效分别为 40.00%、25.33%、34.67%；对菌株 N1-2 引起的根腐病，3 种药剂的防效分别为 35.62%、12.33%、21.92%。50%多菌灵 WP 和 70%甲基硫菌灵 WP 对菌株 QD13-2 和 QZ3 引起的根腐病防效相对较好，50%咯菌腈 WP 对菌株 N12-1 和 N1-2 引起的根腐病防效相对较好。需要进一步优化试验条件或开展田间防治试验，以筛选出对苜蓿镰孢菌根腐病具有良好防效的药剂。

关键词：苜蓿根腐病；镰孢菌；化学防治；药剂筛选

防虫网技术在柑橘园病虫害综合防控中的应用研究

程保平[1],赵弘巍[2],彭埃天[1*],宋晓兵[1],崔一平[1],凌金锋[1]

([1]广东省农科院植物保护研究所/广东省植物保护新技术重点实验室,广州 510640;
[2]南京农业大学植物保护学院,南京 210095)

摘要:柑橘黄龙病和柑橘衰退病对柑橘产业危害巨大,且发病率不断上升,广东省一些产区80%以上橘树出现两种病害的复合感染,加速了树体的衰亡。此二种病害均为虫传病害,其中柑橘黄龙病病原 Candidatus Liberibacter asiaticus 借助虫媒柑橘木虱在田间传播,柑橘衰退病病原 *Citrus tristeza virus*,借助虫媒橘蚜在田间传播。控制虫媒对此二种病害的防控有重要意义。防虫网的应用技术属于一项设施栽培技术,为实现柑橘病虫害的绿色防控开辟了一条新途径。本试验在柑橘园应用60目白色防虫网防治柑橘木虱,橘蚜,每三个月施用一次杀虫剂;再无防虫网的处理组中,每45天施用一次杀虫剂;对照组无防虫网也不施用杀虫剂。最后评价了各处理组的柑橘木虱和橘蚜防效效果;进一步测试了各处理组对柑橘树体内黄龙病菌含量及衰退病毒含量的影响。结果:对柑橘木虱的防效,防虫网+药剂处理组为100%,显著高于药剂处理组的88.6%;对橘蚜的防效,防虫网+药剂处理组为95.68%,显著高于药剂处理组的81.36%。对柑橘黄龙病菌菌量的影响,防虫网+药剂处理组为对照组菌量的27.15%;对柑橘衰退病毒含量的影响,防虫网+药剂处理组为对照组菌量的14.36%。结果表明:防虫网技术可减少药剂使用量50%;对柑橘木虱和橘蚜的防效更佳;还可延缓黄龙病菌和衰退病毒的增殖,该研究可为柑橘病虫害的绿色防控提供一个新的技术储备。

基金项目:广东省重大科技计划专项(2014B020203003);广东省自然科学基金(2014A030310198);广东省现代农业产业技术体系建设专项(2016LM1077);广东省科技计划项目(广东省农作物病虫害绿色防控技术研究开发中心建设);广东省农业标准化项目(粤财农[2016]310号);广东省科技计划重点实验室项目(2014B030301053)

通讯作者:彭埃天(1962-),男,研究员;E-mail:pengait@163.com
第一作者:程保平(1982-),男,助理研究员;E-mail:1093449912@qq.com。

小麦赤霉病籽粒抗性评价指标探究

李长成，龚璇，吉广岩，李磊，李韬*

（扬州大学，扬州 225009）

摘要：小麦赤霉病是主要由禾谷镰刀菌引起的一种世界性病害，该病害的爆发会造成小麦产量和质量的严重下降。赤霉病抗性类型主要分为抗侵染、抗扩展、抗毒素积累和籽粒抗性。前人主要针对扩展抗性进行了大量研究，而对于籽粒抗性的研究较少且结论尚不统一，发表文献中籽粒抗性主要以感染籽粒率（FDK）作为衡量指标，但因其受扩展抗性影响较大且评价标准因人而异导致误差较大。本研究主要探究新的籽粒抗性评价指标来回答是否存在籽粒抗性及如何度量的问题。利用不同发育时期的抗感品种及近等基因系，通过离体试验探究不同发育时期抗感籽粒对禾谷镰刀菌的响应，均匀排列抗感籽粒于 PDA 培养基上并于中央加入直径为 5 mm 禾谷镰刀菌块，定期观测菌落形态及生长速度，结果表明，放置抗病籽粒的菌落生长比放置感病籽粒菌块生长速率快且菌丝茂密，抗病籽粒并不具有抑菌效果，反而促进菌的生长，与对照相比（单独放置菌块不放置籽粒的培养基），放置抗、感籽粒的培养基中的菌丝生长均快于对照菌丝，表明抗感品种的籽粒对禾谷镰刀菌生长均有促进作用，且诱导能力不同。上述结果表明具有扩展抗性品种的籽粒可能并不存在籽粒抗性。

关键词：小麦；赤霉病；籽粒抗性；禾谷镰刀菌

基金项目：国家自然科学基金项目（31771772）；国家重点研发计划“七大作物育种试点专项”子课题（2017YFD0100801）；扬州市重点研发计划（现代农业）（YZ2016035）；江苏省高校优势学科建设工程资助项目（PAPD）

通讯作者：李韬，博士，教授，主要从事小麦赤霉病抗性和小麦硒生物强化遗传研究；E-mail：taoli@yzu.edu.cn

第一作者：李长成，硕士研究生，主要从事小麦赤霉病抗性研究；E-mail：1628725885@qq.com。

玉米抗南方锈病鉴定方法研究及主推品种抗锈病测定

刘彬[1]，胡海平[1]，王琰[2]，张茹琴[1]，王志奎[3]，夏淑春[1*]

([1]青岛农业大学农学与植物保护学院，青岛 266109；
[2]山东省乳山市农业局，青岛 266100；
[3]即墨市农业局，青岛 266201)

摘要：近年来玉米南方锈病在我国一些玉米产区逐年加重，已经严重威胁到玉米生产安全，但对这一病害的控制国内显得准备不足，没有形成有效的防控方法和技术。本研究，采用田间跟踪调查和离体接种方法研究了玉米南方锈病抗性的测定方法和技术，意在探寻实用、快速检测寄主抗锈性和锈病菌致病性变异的技术方法，并借此来认识山东省近年玉米主导品种对南方锈病的抗性及变化情况。结果显示，2015 年玉米锈病在山东省发生最重，调查的 15 个玉米主导品种均发病，发病率为 32.3%～100%，病情指数为 4.8～84.3，品种间差异显著。其中，郑单 958、鲁单 818、济玉 901 和鲁单 9066 四个品种 100%发病，都达到感病级别。2014 年和 2016 年玉米南方锈病在我省相对 2015 年发生轻，平均下降 1～2 个抗性级别，如郑单 958 品种在 2015 年抗性表现为 HS，而 2014 年和 2016 年为 MR，差 2 个级别，浚单 20 品种 2015 年抗性表现为 HS，而 2014 年和 2016 年为 S，但总体上玉米南方锈病在山东省的发生还是呈上升态势。

在调查的 19 个主导品种中，高抗、抗病、中抗、感病和高感品种都有，且差异显著，从 2015 年调查结果看，没有高抗品种，对锈病表现感病和高感的品种偏多，合计占到总数的 2/3，可能是病害流行的主要原因。另外，与黄淮海地区主导玉米品种比较后发现，山东省玉米主导品种也与黄淮海地区的玉米主导品种重复率较高，这些也可能是近年来本地区玉米南方锈病逐年加重甚至流行的另一个重要原因，建议相关机构关注玉米锈病的危害性，尽快培育并推广可利用的抗病品种，建立有效的玉米南方锈病防控技术，控制该病害的进一步危害。

玉米南方锈病抗性室内测定结果表明：玉米南方锈病菌最适宜的孢子萌发湿度是在水膜中，相对湿度低于 90%很少萌发；夏孢子萌发的最适宜温度是 26～28℃，低于 14℃和高于 34℃几乎不萌发；温湿度对玉米南方锈病菌夏孢子的萌发有显著影响。在 27℃气温条件下，发病率和病情指数是随生育期增进而降低，尤其是病情指数表现更显著。由此可以看出玉米不同生育期抗锈病不同，苗期最感病。对山东省近 3 年主导玉米品种抗南方锈病室内测定结

基金项目：山东省自然科学基金项目(ZR2011CL005)；山东省“泰山学者”建设工程专项经费资助(BS2009NY040)
通讯作者：夏淑春(1964-　)，女，辽宁锦州人，副教授，主要从事植物病理学研究；E-mail：songxy@qau.edu.cn
第一作者：刘彬(1993-　)，男，山东省青岛人，植物保护专业本科学生；E-mail：dedj@qau.edu.cn。

果显示：与田间抗病性调查结果基本一致。郑单 958、浚单 20 和鲁单 818 等品种仍然是高感玉米南方锈病，登海 605、金海 5 号、聊玉 22 抗病，而不是高抗；只有 2016 年刚刚入选主导品种的华良 78、华盛 801 是高抗。说明采用苗期喷雾接种离体叶片的方法可用于玉米植株对南方锈病抗性测定。

木质素合成途径关键酶 COMT 在调控玉米抗病防卫反应中的作用

刘孟洁，孙扬，陈洁，栾庆玲，刘潇颖，王梅，王官锋*

（山东大学生命科学学院，济南 250100）

摘要：*Rp1-D21* 是从玉米中得到的一个自主免疫突变体，是由相似性很高的两个 CC-NB-LRR 类抗病(R)基因 *Rp1-D* 和 *Rp1-dp2* 通过分子内重组而产生。Rp1-D21 能够在玉米和烟草中诱导自主超敏防卫反应(HR)，因此已被成功用于研究调控 Rp1 介导 HR 的关键基因的功能鉴定。木质素在植物发育和抗病中发挥重要功能，已有多种木质素合成途径关键酶被报道参与植物对病原菌的防御响应。我们前期研究发现，木质素合成途径的关键酶 hydroxycinnamoyltransferase（HCT）和 caffeoyl-CoA O-methyltransferase（CCoAOMT）通过与 Rp1-D21 相互作用调控 HR 的表型。caffeic acid O-methyltransferase(COMT)是木质素合成途径中的另一关键酶，玉米 COMT 突变产生棕色中脉突变体(*brown midrib 3*，*bm3*)，然而 COMT 在玉米中调控防卫反应的作用机理尚不清楚。我们详细研究了玉米 COMT 在 Rp1-D21 介导的 HR 中的调控功能。首先通过农杆菌介导的瞬时表达系统将 Rp1-D21 与 COMT 及其 2 个同源蛋白(COMT2、COMT3)在烟草中分别共表达，发现 COMT2、COMT3 对 Rp1-D21 介导的 HR 有抑制功能。COMT2、COMT3 还能抑制 Rp1-D21 的信号功能域 coiled coli(CC)介导的 HR。通过 Co-IP 实验证明，COMT2、COMT3 能够与 CC 结构域发生相互作用。进一步研究发现 COMT2、COMT3 能够与 HCT 和 CCoAOMT 相互作用。综合以上结果，我们推测木质素合成途径多个关键酶通过形成一个大的复合物参与调控 Rp1-D21 介导的 HR。本研究将初步阐明玉米 COMTs 在抗病蛋白 Rp1 介导的防卫反应中的调控功能。

关键词：玉米；超敏防卫反应；Rp1-D21；COMT

通讯作者：王官锋，教授，玉米抗病分子机理研究；Tel：(+86531) 88364532，E-mail：gfwang@sdu.edu.cn

第一作者：刘孟洁，博士，玉米抗病分子机理研究；E-mail：mjliu@sdu.edu.cn。

辣椒资源对辣椒脉斑驳病毒抗性水平的鉴定

汤亚飞[1,2]，何自福[1,2*]，佘小漫[1]，蓝国兵[1]，于琳[1]，邓铭光[1]

([1]广东省农业科学院植物保护研究所，广州 510640；
[2]广东省植物保护新技术重点实验室，广州 510640)

摘要：辣椒脉斑驳病毒（*Chilli veinal mottle virus*，ChiVMV）是马铃薯Y病毒科（*Potyviridae*）马铃薯Y病毒属（*Potyvirus*）成员，为一种（+）ssRNA病毒，主要依靠蚜虫传播，可以通过病株汁液和机械接触等方式传播，但是不能通过种子传播。自1979年在马来西亚首次报道ChiVMV以来，世界上已有多个国家和地区发现该病毒，包括亚洲的韩国、马来西亚、泰国、印度、中国及非洲地区。在我国，目前已在台湾、云南、四川、陕西、海南等多个省市发现了该病毒。2013—2016年，利用sRNA深度测序和RT-PCR相结合方法，本团队对为害广东省辣椒的病毒种类进行了检测与鉴定，发现ChiVMV检出率达到30%以上，是为害广东辣椒的优势病毒。辣椒是广东主要冬种蔬菜之一，年种植面积超过100万亩。病毒病是为害冬种辣椒的主要病害之一，每年均造成不同程度损失。为此，本团队收集了16份辣椒资源，采用室内人工摩擦接种法，评价了这些辣椒资源对ChiVMV广东分离物的抗性水平。结果显示，16份辣椒资源对ChiVMV广东分离物的抗性差异明显，其中免疫1份、高抗5份、抗病1份、中抗1份、感病4份和高感4份。这些结果为辣椒抗病育种提供了科学依据，育种专家可以利用表现为高抗或免疫水平的资源培育出抗ChiVMV的辣椒品种，来满足辣椒生产的需求。

关键词：辣椒脉斑驳病毒；辣椒；抗性鉴定

基金项目：国家公益性行业（农业）科研专项（201303028）；广东省科技计划项目（2015A020209070）

通讯作者：何自福，研究员，主要从事蔬菜病理学研究；E-mail：hezf@gdppri.com

第一作者：汤亚飞，副研究员，主要从事植物病毒学研究。

利用大豆胞囊线虫 HG 类型辅助进行抗性品种培育

陈井生[1,2],马兰[1],李建英[1],于吉东[1],
吴耀坤[1],周长军[1],李肖白[1],田中艳[1]
([1]黑龙江省农业科学院大庆分院,大庆 163316;
[2]沈阳农业大学植物保护学院,沈阳 110866)

摘要:黑龙江省是我国非转基因大豆的主要产区,2017 年大豆播种面积接近 5000 万亩,约占全国的三分之一。大豆胞囊线虫(soybean cyst nematode,SCN)在黑龙江省大豆产区发生普遍,是生产中大豆产量的重要限制因子。大豆胞囊线虫在地下危害而且具有隐蔽性,通常导致产量损失在 30%以下,地上植株没有明显的症状。植物病理学家和育种家们通过选育抗性品种用以防治该病害,使得该病害得到了有效控制。近年来,由于大豆胞囊线虫具有遗传多样性和生产上抗线虫大豆品种数量少、抗性来源单一、抗源不清晰等因素影响了黑龙江省抗线虫大豆品种推广应用,严重影响了我国大豆商业化育种进程。HG 类型鉴定模式能够有效反应大豆胞囊线虫的寄生能力,国际上大豆育种家普遍用以划分不同大豆胞囊线虫群体对大豆的致病性,被广泛应用于抗线虫大豆育种工作中。本文对黑龙江大豆主要产区大豆田 62 份土样进行 HG 类型鉴定,共鉴定出 11 个 HG 类型,其中 HG 0、1.2.3.5.7、1.2.3.7、1.3.4.7、2、2.5.7、2.7、6、6.7 等 9 个 HG 类型为黑龙江省首次报道。根据 HG 类型鉴定结果进行了黑龙江省抗线虫大豆品种改良种质库构建,利用 PI548402(Peking)、PI437654(Hartwig)、PI88788、PI209332 以及五寨黑豆、灰皮支黑豆等不同抗源与当地主栽品种进行杂交,选育具有不同抗性来源的抗病品种。杂交后代材料的低世代,在大豆胞囊线虫病圃进行抗线虫鉴定、选择;高世代进行大豆胞囊线虫盆栽鉴定及品质跟踪分析,定向选择,获得到一些不同抗性来源的新种质。这些种质可作为大豆抗胞囊线虫育种的亲本材料,将对推动大豆生产起到一定作用。

关键词:大豆;大豆胞囊线虫;HG 类型;抗线虫育种

基金项目:现代农业产业技术体系建设专项(CARS-004-CES07);国家自然科学基金重点项目(31330063);大庆市科技局指导性项目(ZD-2016-121)

第一作者:陈井生(1982-),男,副研究员,博士,主要从事植物线虫学和大豆抗线虫育种研究;E-mail:jingsheng6673182@163.com。

一种可延缓柑橘黄龙病危害的抗病栽培方法研究

程保平[1],赵弘巍[2],彭埃天[1*],崔一平[1],宋晓兵[1],凌金锋[1]

([1]广东省农科院植物保护研究所/广东省植物保护新技术重点实验室,广州 510640;
[2]南京农业大学植物保护学院,南京 210095)

摘要:柑橘黄龙病对世界柑橘产业造成巨大威胁,而目前尚无治愈该病的方法。本研究在100%感染黄龙病的橘园里开展了抗病栽培试验,并在处理后第0、5和10个月后记录病树内黄龙病菌的含量变化,以明确抗病栽培是否可延缓黄龙病菌的增殖。本论文所用的抗病栽培方法主要包括:田间少量多施有机肥,多次进行有机肥水滴灌。结果表明:使用抗病栽培处理5和10个月后,病树中的黄龙病菌含量比对照分别显著降低了36.49%和31.12%,土壤中的有机质增加了8.07%,土壤细菌种群数量增加28.81%,土壤放线菌种群数量增加15.96%;土壤真菌种群数量增加19.07%,病树的总产量增加28.91%,平均果实重量增加18.74%,果实可溶性固体含量增加17.07%,可溶性糖增加了23.35%,总酸降低了19.61%。结果表明:本研究所用抗病栽培可以改善土壤性质、延缓黄龙病菌增殖,并可改善病树的产量与质量。

关键词:柑橘黄龙病;抗病栽培;病菌滴度;果实产量;果实质量

基金项目:广东省重大科技计划专项(2014B020203003);广东省自然科学基金(2014A030310198);广东省现代农业产业技术体系建设专项(2016LM1077);广东省科技计划项目(广东省农作物病虫害绿色防控技术研究开发中心建设);广东省农业标准化项目(粤财农[2016]310号);广东省科技计划重点实验室项目(2014B030301053)

通讯作者:彭埃天(1962-　),男,研究员;E-mail:pengait@163.com
第一作者:程保平(1982-　),男,助理研究员;E-mail:1093449912@qq.com。

拟南芥先天免疫增强突变体 *aggie102* 中目的基因的图位克隆与表型分析

孙炜[1],齐婷[1],刘若西[1],何平[2],单丽波[3],孙文献[1],崔福浩[1*]

([1]中国农业大学植物保护学院,农业部作物有害生物监测与绿色防控重点实验室,北京 100193;
[2]美国 Texas A & M University 生物化学与生物物理学系,77840;
[3]美国 Texas A & M University 植物病理学与微生物学系,77840)

摘要:植物细胞表面的免疫受体能识别病原微生物的相关分子模式(PAMPs),进而引起植物产生活性氧爆发、蛋白激酶级联途径激活、基因转录重排等一系列防卫反应,但目前对植物先天免疫调控网络的认识仍不清楚。利用前期建立的 *pFRK1*::*luciferase* 转基因拟南芥 EMS 突变体库,我们筛选到几十个先天免疫信号发生显著改变的突变体,*aggie102* 是其中一个免疫信号显著增强的突变体。通过粗定位和精细定位,目前已将突变基因的位置锁定在第 4 条染色体上 75 kb 大小的区域内。经检测,细菌鞭毛蛋白 flg22 在 *aggie102* 中能诱导更多的活性氧爆发。有意思的是,4~5 周大小的 *aggie102* 植株在无任何处理的情况下免疫信号仍能被部分激活,而向叶片注射 flg22 后则能进一步激活 *pFRK1*::*LUC* 的活性。推测目的基因在拟南芥先天免疫信号途径中起到负调控作用。另外,*aggie102* 植株也表现出开花延迟、结实率低、叶片沿两侧向下卷曲等生长发育表型。抗病性等表型的测定将有助于阐明目的基因在植物先天免疫中的重要作用。同时,我们也在结合图位克隆和全基因组测序技术来鉴定其他 *aggie* 突变体中的目的基因。*Aggie* 基因的鉴定和功能解析将有助于深入理解植物免疫信号调控网络,可望为作物抗病育种提供重要的遗传资源。

关键词:拟南芥;先天免疫;信号传导;突变体筛选

基金项目:中国农业大学基本科研业务费(2017QC070)
通讯作者:崔福浩,副教授,主要从事植物与病原细菌、真菌互作;E-mail:cuifuhao@163.com
第一作者:孙炜,硕士研究生,研究方向:植物与病原细菌的分子互作;E-mail:1259502394@qq.com。

剑麻类 *Bsr-d1* 基因的鉴定与表达分析

黄兴，梁艳琼，习金根，郑金龙，贺春萍，吴伟怀，李锐，易克贤*

（中国热带农业科学院环境与植物保护研究所，
农业部热带农林有害生物入侵检测与控制重点开放实验室，
海南省热带农业有害生物检测监控重点实验室，海口 571101）

摘要：*Bsr-d1* 在水稻中编码 C_2H_2（锌指蛋白类）转录因子，该基因启动子的一个关键碱基变异可减弱 H_2O_2 降解从而提高水稻对稻瘟菌的免疫反应及抗病性。这一水稻稻瘟病广谱抗病机制的发现为其他作物新型抗病机理的研究提供了重要借鉴。本研究以水稻 *Bsr-d1* 基因为检索序列，对已公布的剑麻转录组进行 blast 检索后获得 1 条同源的剑麻基因序列。将该序列在 NCBI 数据库中比对后，发现其包含完整的编码序列，并将其命名为 *AsBsr-d1-like1*。该基因开放读码框具有 528 个碱基，编码 176 个氨基酸，具有 2 个 C_2H_2 保守功能域。系统进化分析表明该基因与芦笋 *ZAT6-like* 基因亲缘关系较近。对斑马纹病病原菌烟草疫霉入侵前后的剑麻叶片进行转录组测序，结果发现该基因在病原入侵后表达水平显著上调。荧光定量 PCR 检测结果也显示该基因在病原入侵后上调，与转录组测序结果一致。此外，转录组测序结果中还发现多条超氧化物歧化酶（POD）基因显著上调，表明 *AsBsr-d1-like1* 基因的上调可能与 POD 基因的上调存在一定关系。剑麻类 *Bsr-d1* 基因的发现表明剑麻中可能存在类似于水稻 *Bsr-d1* 介导的广谱抗病机制，上述结果将为剑麻斑马纹病抗病性研究提供新的方向。

关键词：剑麻；类 *Bsr-d1* 基因；表达

通讯作者：易克贤，男，博士，研究员，研究方向：分子抗性育种；E-mail：yikexian@126.com
第一作者：黄兴（1988- ），男，博士，助理研究员，研究方向：分子抗性育种；E-mail：huangxingcatas@126.com。

植物病害系统监测与病害精准防治

王海光*

(中国农业大学植物病理学系,北京 100193)

摘要:植物病害是农业生产的重要限制因素,是造成农产品产量和品质降低的重要原因,也是影响粮食安全的重要因子。做好植物病害防治意义重大,具有重要的经济效益、生态效益和社会效益。社会发展对病害防治提出更高要求,既要将病害控制在经济损害允许水平之下,又要降低对环境的破坏,特别是要降低防治中农药的使用量和残留。因此,开展植物病害的精准防治具有十分的必要性。在植物病害调查、监测和预测基础上,在一定时间范围或空间范围内通过单一措施或综合利用多种措施对病害有针对性的防治,可称为植物病害精准防治(precision control of plant disease)。病害精准防治是精准农业的重要组成部分和重要体现,是建立在对病害发生位置、发生时间、发生过程或主要致病基因等信息精确掌握的基础上的。从不同角度考虑,"精准防治"可以有不同内涵。从防治对象方面来说,明确病害种类和病原,瞄准病害发生的关键原因,甚至主要致病基因、靶向基因或位点,做到对症防治或靶向防治,可谓精准;从空间角度来说,明确病害发生区域、危害植物部位,进行定位防治,可谓精准;从时间上来说,明确病害发生的时间范围、发生时期以及病害所处发展阶段,瞄准病害发生关键时期采取防治措施,可谓精准,特别是在病害发生最薄弱时期采取措施,会收到事半功倍的效果;从防治措施方面来说,明确防治措施的靶向性,做到防治有度,既节省人力和物力,又达到有效防治效果,亦可谓精准,特别是利用化学农药防治时,把握好施药时机、施药量、施药方式等,借助GPS定位技术、病害识别技术等精准施药,对于"减药"和提高防治效果尤为重要。"精准"是相对的,与对病害的认识水平、防治要求以及所具有的防治条件等有关。要做到植物病害精准防治,需要掌握有关病害系统各组分的状态和关系的信息。因此,植物病害系统监测(plant pathosystem monitoring, monitoring of plant pathosystem 或 monitoring of plant disease system)是病害精准防治的基础和前提,其是对由病原和寄主在一定环境中通过相互作用构成的开放、动态的生态系统的状态和变化进行的观察和记录,监测的具体对象包括病原物、寄主、病害、环境、人类干预活动等。植物病害系统监测在空间上可分为洲际水平、国家水平、地区水平、田块水平……,在时间上可分为多年、单年、单个生长季节、单个病程等不同水平。病害精准防治很大程度上取决于植物病害系统监测的精确度和准确度。若要做到病害精准防治,做好病害系统的监测尤为重要。近年来,植物病害系统监测的技术和方法得到了迅速发展,如分子生物学技术、遥感技术、计算机视觉技术以及综合多种技术的物联网技术等极大地提高了监

基金项目:国家自然科学基金项目(31471726)

通讯作者:王海光,副教授,主要从事植物病害流行学和宏观植物病理学研究;E-mail: wanghaiguang@cau.edu.cn。

测水平。不但要重视分子生物学方法和技术的研究和应用,更要重视植物病害流行学和宏观植物病理学的研究。避免只重视微观方面的研究,而忽视宏观方面的研究。既要有针对性地进行抗病基因的筛选和品种选育,做好抗病品种的合理利用,又要探索和发展植物病害和病原快速诊断技术和早期检测技术、病害系统信息获取技术和处理技术等。重视综合信息获取、传输、处理和判别技术的物联网技术的研发和应用,加强植保信息化建设,依靠大数据、云计算、机械学习等技术,发挥人工智能的作用,做好植物病害系统的自动化、数字化、智能化监测,从而为病害的精准防治提供精准信息和支撑。

关键词:植物病害系统;监测;病害防治;精准防治;植保信息化